P9-CQY-820

The American Immigration Collection

Anthracite Coal Communities

PETER ROBERTS

Arno Press and The New York Times

NEW YORK 1970

Reprint Edition 1970 by Arno Press Inc.

Reprinted from a copy in
The Framingham Town Library

LC# 75-129410

ISBN 0-405-00564-4

The American Immigration Collection—Series II

ISBN for complete set 0-405-00543-1

Manufactured in the United States of America

ANTHRACITE COAL COMMUNITIES

Anthracite Coal Communities

A Study of the
Demography, the Social, Educational and Moral Life
of the Anthracite Regions

By

Peter Roberts, Ph.D.

Author of "Anthracite Coal Industry"

New York
THE MACMILLAN COMPANY
LONDON: MACMILLAN & CO., Ltd.
1904

Set up and published, January, 1904

Press of
The New Era Printing Company,
Lancaster, Pa.

PREFACE.

When the strike of 1900 was settled, all who knew the situation at first hand felt that the settlement was only an armistice, that the real conflict between capital and labor in the anthracite regions was yet to come. The great strike of 1902 came and with it a harvest of misery, privation and crime. It cost us over $100,000,000 and wrought moral ruin the extent of which none can estimate. The outcome of the conflict — the interference of the President and the appointment of a commission — was not dreamt of by the most sanguine advocates of the rights of labor. For over four months, the Coal Strike Commission inquired into the "economic, domestic, scholastic and religious phases" of the mine workers' lives. It examined 558 witnesses and most of the testimony was eagerly read by an interested public. During the conflict fundamental questions relative to industrial and social relations were raised. Men of national fame, discussing the issues involved, astonished their most intimate friends by proposing solutions so radical as to be little short of a complete subversion of our industrial system. In the sessions of the Commission, all attempts to limit the scope of the inquiry to the industrial questions which precipitated the conflict were vain. To 80 per cent. of mine workers the question of wages meant their whole living and the Commission was forced to listen to the story of these people's life in all its phases. Never before, in any industrial dispute, was it more clearly seen that the students of the industrial and social problems are laboring for identical ends, and that the reformers of the industrial and social world are fighting under the same banner.

The anthracite employees, since the close of the great strike, have had a year of unparalleled prosperity. From November 1, 1902, to November 1, 1903, over 62,000,000 tons of coal were mined. Both employers and employees have prospered, but industrial prosperity is not synonymous with social progress, and our greatest danger arises to-day from the tendency to regard "the Belly and its adjuncts as the great Reality." Should there not be means instituted to counteract this tendency?

I have given the facts relative to the economic life of our people in "The Anthracite Coal Industry," published in the fall of 1901. The object of the present volume is to give the facts relative to the social and moral life of the anthracite mine employees. Let the reader deeply ponder the facts given in the following pages and ask if considerations of patriotism demand not the coöperation of all citizens for the future peace and progress of anthracite communities.

I desire to express thanks for many suggestions and corrections to Professors W. G. Sumner and Henry W. Farnam, of Yale University, to Professor Katherine Coman, of Wellesley College, and especially to Dr. H. T. Newcomb, of Washington, D. C., who carefully read the manuscript and made many criticisms and suggestions which assisted me in making improvements in the book.

P. R.

MAHANOY CITY, PA.,
January, 1904.

CONTENTS.

ILLUSTRATIONS.

BIBLIOGRAPHY.

The statistics were gleaned from the following works:

State reports:

Bureau of Mines of Pennsylvania: Reports of, 1870–1902. 32 vols.

Board of Health of Pennsylvania: Reports of, 1900–1902. 4 vols.
    Circulars and Laws of.

Commissioner of Banking: Reports of.
    Banks and Trust Companies: 1900–1901. 2 vols.
    Building and Loan Associations: 1900–1901. 2 vols.

Commission of Inquiry into the Condition of the Insane in Hospitals: Report of (1902).

Factory Inspectors: Reports of, 1897, 1899, 1900 and 1901. 4 vols.

Industrial Statistics, 1890, 1900 and 1902. 3 vols.

Legislative Handbook: Smull's.

Public Charities of Pennsylvania, Reports of, 1891, 1900 and 1901. 3 vols.

School Laws of Pennsylvania (1902).

Superintendent of Public Instruction: Reports of, 1899, 1900, 1901 and 1902. 4 vols.

Other sources:

Anthracite Coal Strike Commission, Report of (1903).

Ayers, N. W. Newspaper Annual.

Building and Loan Associations. Commissioner of Labor's Report of (1893).

Catholic Directory (1902).

Census returns for 1860, 1870, 1880, 1890 and 1900.

Controller of the Currency. Report of (1901).

Court Houses of Anthracite Counties.
    Dockets of Births and Marriages.
    Assessors' Returns.
    Criminal Dockets.
    License Dockets.
    Naturalization Dockets, etc.

Delaware and Hudson Coal Company. Reports of Accidental Fund from 1887–1901.

Hospitals (ten) in Anthracite Regions, Annual Reports of.

Justices of the Peace in Mining Towns, Dockets of.

Labor Statistics, New York, 1899.

Leading Denominations in the Coal Fields, Annual Reports of.

Local Boards of Health, Dockets of.

Mining Municipalities, Reports of.

Philadelphia and Reading Coal and Iron Company. Reports of Accidental Fund, 1888–1902.

Rowell, Geo. P. American Newspaper Directory.

We derived help from the following books:
ASHLEY, W. J. "English Economic History."
BAGEHOT, WALTER. "Physics and Politics."
BLATCHFORD, ROBERT. "Merrie England."
BOISE, H. M. "Prisoners and Paupers" and "Science of Penology."
BROOKS, J. G. "Social Unrest."
CARLYLE, THOMAS. Works, 12 vols.
CLARK, J. B. "Distribution of Wealth."
COMMITTEE OF FIFTY. "Economic Aspects of the Liquor Problem."
CONE, ORELLO. "Rich and Poor."
DONISTHROPE, WM. "Individualism."
ELLIS, HAVELOCK. "The Criminal," and "Man and Woman."
ENCYCLOPEDIA BRITTANICA." (Ninth edition) Articles "Sclav," "Russia," "Poland," "Letts," etc.
FORBUSH, W. B. "The Boy Problem."
GEDDES AND THOMPSON. "The Evolution of Sex" (1902).
GIDDINGS, T. H. "Principles of Sociology."
GIDE, CHARLES. "Political Economy."
GILMAN, N. P. "Socialism and the American Spirit."
GRAHAM, WM. "Socialism New and Old."
GRASSERIE, RAOUL DE LA. "Des Religions Comparees."
GUYAU, M. "L'Irreligion De L'Avenir."
HADLEY, A. T. "Economics."
HARNACK, ADOLF. "What is Christianity?'
HARRIS, GEORGE. "Moral Evolution" and "Inequality and Progress."
HENDERSON, C. R. "The Social Spirit in America."
HUXLEY, T. H. "Elements of Physiology."
HOPPE, DR. HUGO. "Die Thatsachen über den Alkohol" (1901).
LETOURNEAU. "Le Mariage."
LILIENFELD, P. DE. "La Pathologie Sociale" (1896).
LORIA, A. "Les Bases Economiques de la Constitution Sociale," and "Principes Sociaux Contemporains."
LIPPERT, JULIUS. "Kulturgeschichte." 2 vols. (1886).
MARX, CARL. "Capital."
MARSHALL, ALFRED. "Principles of Economics" (1898).
MATTHEWS, SHALER. "The Social Teaching of Jesus."
MAYO-SMITH. "Statistics of Sociology."
NITTI, F. S. "La Population et le Systeme Social" (1897).
OLIVER, THOMAS, M.D. "Dangerous Trades."
PAULSEN, FRIEDRICH. "A System of Ethics."
PEABODY, F. G. "Jesus Christ and the Social Question."
PLOSS-BARTELS. "Das Weib." 2 vols. (1902).
PURDON'S "Digest of the Laws of Pennsylvania."
RAE, JOHN. "Contemporary Socialism."
RANKE, DR. JOHANNES. "Der Mensche." 2 vols. (1894).
ROBERTS, PETER. "Anthracite Coal Industry."
RUSKIN, JOHN. Works. 14 vols.
SMALL AND VINCENT. "An Introduction to the Study of Society."

SMITH, ADAM. "Wealth of Nations."
SMYTHE, NEWMAN. "Christian Ethics."
SPENCER, HERBERT. "Education."
"A Study of Sociology."
"First Principles."
"Principles of Sociology." 3 vols.
"Principles of Biology." 2 vols.
"Principles of Psychology." 2 vols.
"Data of Ethics."
SUMNER, W. G. "What Social Classes Owe to Each Other."
"Problems in Political Economy."
TURGOT. "The Formation and the Distribution of Riches."
WARD, L. F. "Dynamic Sociology." 2 vols.
"Psychical Factors of Civilization."
"Pure Sociology."
WEIMER, A. B. "Laws Relating to the Mining of Coal in Pennsylvania."
WRIGHT, CARROLL D. "Ethical Considerations in Distribution of Wealth."
VIGNES, J. B. M. "La Science Social." 2 vols.
VIRTUE, Dr. "Anthracite Mine Laborers." Bulletin of Department of Labor (1897).
ZANTEN, J. H. VAN. "Die Arbeiterschutzgesetzgebung in den Europaischen Landern" (1902).

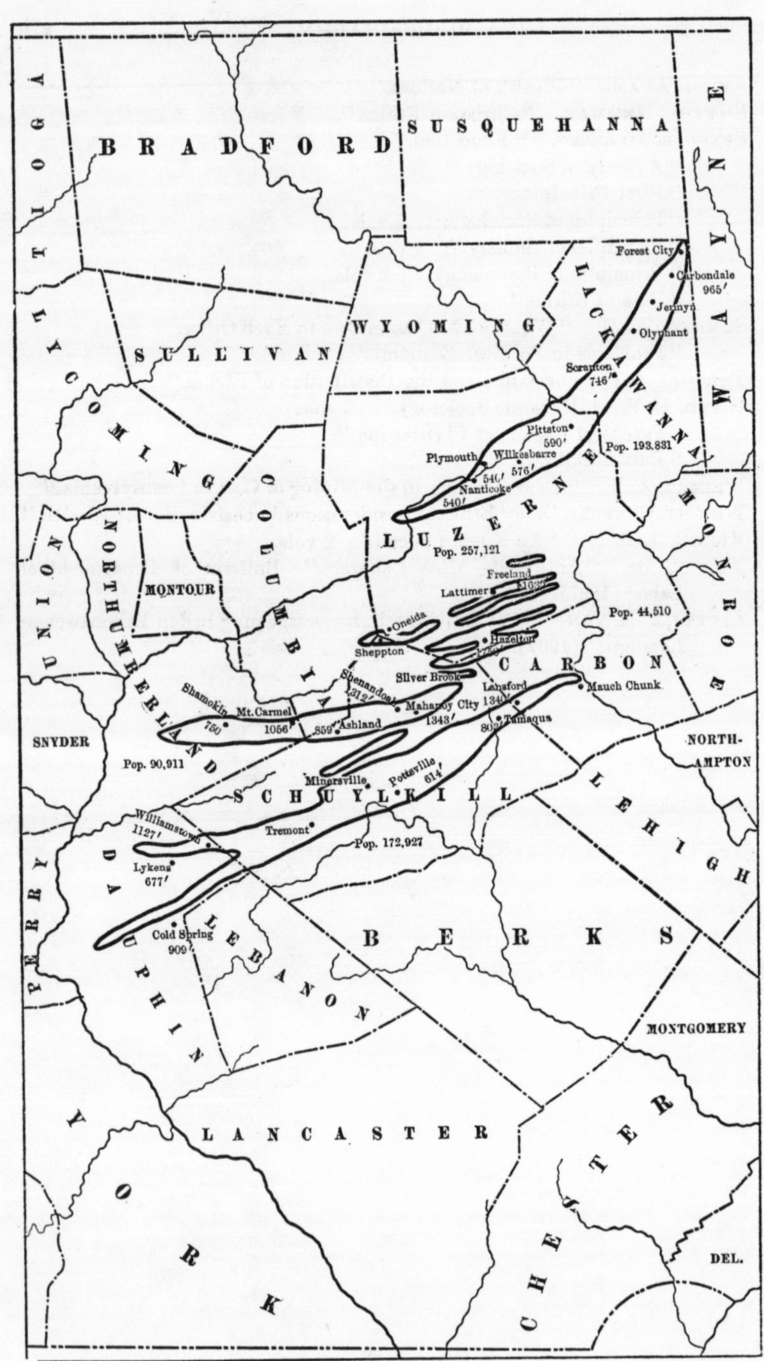

Map of Coal Fields Giving Cities and Principal Towns with Elevation above Sea-level.

ANTHRACITE COAL COMMUNITIES.

CHAPTER I.

OUR MINING POPULATION.

1. The Territory in which Anthracite Miners Live. 2. Immigration and Social Capillarity. 3. The Twenty-six Peoples now Residing Here.

The Territory.

The anthracite coal fields are located in the northeastern portion of the State of Pennsylvania. The various coal basins, outlined on the accompanying map, appear as so many islands upon which representatives of twenty-six different nationalities have settled who gain their means of subsistence by exploiting these rich coal fields. The density of population of the counties of Lackawanna, Luzerne and Schuylkill as compared with adjoining counties is due to the mining industry, and when their supply of coal shall have been exhausted, these will be chiefly populated by agriculturalists as was the case before their coal deposits were discovered.

The Appalachian valley, of which these coal fields form a part, consists of alternate and parallel, narrow sinuous ridges and valleys which have been greatly folded, faulted and eroded. The ridges consist of sandstone and quartzite. At the outcrops of the coal basins the soil is red, because of the red shale which forms the bed of the anthracite coal fields. The valleys lying between these ridges consist of soft and soluble limestone and the diluvial deposits of rivers and creeks, which now are narrowed down to small beds as compared with the expanse they once occupied.

A line drawn from the southeastern point of Susquehanna county passing through Lackawanna and Luzerne counties to the extreme limit of the Wyoming basin at Shickshinny, then down south through Columbia county to Centralia, and again west and south around the off-shoots of the Schuylkill and Pottsville basins passing through the counties of Northumberland, Dauphin and Lebanon; then northeast through Schuylkill and Carbon counties to Mauch Chunk, and north again to the point of departure, will enclose the total area of the anthracite coal fields, which is 1,700 square miles. The coal deposits, however, occupy only 480 square miles.

The section lies between 75 and 77 degrees longitude and 40 and 42 degrees latitude. Thus the coal fields lie 5 degrees nearer the equator than does the territory occupied by the Sclav races of Europe which now form an essential part of this population, while it is 10 degrees nearer than the British Isles. The meteorological observations of the weather bureau as observed at Wilkesbarre give an average mean temperature during the last fifteen years of 50.7 degrees Fahrenheit: the highest being 101 degrees and the lowest −9 degrees. The average annual precipitation covering the same number of years was 35.02 inches; the maximum in any one month being 4.49 inches and the minimum 1.29. An annual average of 138 days of sunshine was enjoyed during the fifteen years, while 165 days were cloudy and 62 partly cloudy. Thus from the standpoint of physical environment, the temperature, precipitation and atmospheric conditions are most favorable to health.

The anthracite basins are embedded in the bosom of the hills, though the towns and villages vary considerably in their elevation above sea-level. In the Wyoming valley at Wilkesbarre, where we have the lowest elevation in the coal fields, the city is 575 feet above sea-level. As we ascend the bed of the Susquehanna to Pittston, we find there an elevation of about 590 feet. Scranton is 746 feet, and Carbondale stands 965 feet above the level of the sea. The highest elevation is near Hazleton, which is 1,750 feet above sea-level; from the Hazle-

The Head of a Modern Shaft Showing Tanks for Hoisting Water from a Depth of 800 Feet.

ton mountains we descend to Mahanoy City and Shenandoah where we have 1,343 and 1,312 feet respectively; following the Lykens Valley basin as far as Shamokin we find there an elevation of 750 feet. Passing into the Pottsville basin we find at Lansford an elevation of 1,340 feet and at Tamaqua 803 feet. Ashland is 859 feet above the level of the sea, and Pottsville 614 feet. Following the outcrops from here into Dauphin and Lebanon counties we have an elevation of 1,127 and 909 feet at Williamstown and Cold Spring respectively.

The territory is drained by two majestic rivers, the Delaware on the east and the Susquehanna on the west. The Lackawanna river flows over the northern section of the Wyoming basin and joins the Susquehanna at Pittston. Entering the coal fields at this point the Susquehanna river flows over the remaining section of the northern basin as far as Shickshinny. Here it bends south and sweeps around the southern coal fields, draining the northern part of both Luzerne and Schuylkill counties. The southern portion of these two counties, in which lie the Middle and Southern coal fields, is drained by the Lehigh and Schuylkill rivers, both of which flow into the Delaware.

The mining industry has effected a great change in the physical features of this section of the state of Pennsylvania. Many immigrants, still residing in the coal fields, well remember these mountain streams filled with fishes, which afforded ample sport to the pioneer settler. The hills were covered with virgin forests. The first industry which thrived in the region was that of lumbering, and many localities still preserve the memories of lumber-camps and saw-mills which flourished in the last century. The product of these saw-mills was generally hauled to the banks of creeks and rivers and, being firmly bound into rafts, was floated down the streams and brought to market. The creeks and rivers were then a source of joy and blessing to men. Pebbles of sandstone and quartzite glistened on the river banks, and gravel beds of pristine brightness were seen through the transparent streams. Fish played in these crystal waters which murmured a sweet lullaby to the Indian, the pioneer agriculturalists and lumbermen, weary with the day's toil. A great change has come over this charming landscape. The hills have been stripped of their forests and the major part of the graceful trees which once bent to the storm now rot in the caves of the earth, where, for a time, they afforded protection to mine employees engaged in digging coal. The climate of these regions has been considerably modified. The long winters and heavy snow falls, which the pioneer settlers regularly expected, no longer prevail. The territory which then was sparsely settled is now studded with villages, towns or cities. But in nothing is the change so marked as in the character of these mountain streams. When the woodland heights formed reservoirs for the melting snows, the creeks and rivers were seldom flooded, and rarely were they known to dry up in summer. Now the rain and snow have no natural reservoirs. A heavy storm immediately swells the streams. Every storm means a flood. The waters soon pass to the mighty deep and the land is again thirsty. When the winter snows melt, the creeks and rivers overflow their banks, and the rapid rush of many waters into the Susquehanna and Delaware results in inundations which cause an incomputable annual loss to farms and towns along the banks of these streams.

This, however, is not all the change effected in the character of these waterways. The mining industry perfects the work of destruction. A casual trip in railroad cars over the Hazleton mountains and down through Schuylkill county will show the extent of the damage done to the surface by mining operations. The Mammoth vein has been largely worked and abandoned and the surface has caved in, so that everywhere the ragged sides of these huge depressions appear. This disfigurement is common throughout the coal fields though not to the same extent in the Northern as in the Middle and Southern. The result is that creeks, which once carried away the surface water from these hills, are now dry. The water is drained by these surface depressions. It enters the mines and must again be pumped to the surface. During the floods in December, 1901, a foreman in one of the Reading shafts in Schuylkill county, on a tour of inspection through the mines, was met by a flood two feet deep rushing along the gangway. The water-ways which once were on the surface are there no more — streams in the depths of collieries have taken their place.

The quality of the creeks and rivers is also very different. The waters of nearly 400 collieries impregnated with sulphur, flowing into the creeks and rivers have killed the fish. In addition to this, the necessity in recent years of washing the coal before sending it to market as well as the erection of washeries for the overhauling of the culm heaps, adds to every available stream a quantity of foul water laden with coal dust which turns the creeks and rivers into a mass of black flowing stuff that is a curse to all forms of organic life. Then in these valleys the culm heaps — the accumulated refuse of nearly eighty years of mining — are located. When heavy rains fall, the waters carry a certain quantity of this culm into the creeks and rivers, so that both bank and bed are to-day covered with coal dust, and the waters of the Susquehanna and the Delaware, which once were a joy to men, have become a curse. To this defilement of the streams another still more poisonous defilement is added. The towns and cities located on the banks of these creeks and rivers turn the contents of their sewers into them. This pollution imperils the health of mining communities themselves. In some regions the streams are little better than open sewers contaminating the air with poisonous gases. The occasional flooding of these water-ways above referred to is a blessing, for then the accumulated filth that poisons the atmosphere is swept away. In summer time, when the refuse of towns, the garbage of cities, the foul contents of sewers, and the black sulphurous water of mines and washeries, defile the streams, the dismal sight and the foul stench oppress the hearts and defile the souls of men. We pay the price of our carelessness, for annually scores of innocents perish in the homes of the people. Judge Thayer, of Philadelphia, said "to pollute a stream is to maintain a common nuisance. It is not only a public injury, but it is a crime, a crime for which those who perpetrate it are answerable in a tribunal of criminal jurisdiction." The law of the State, however, has favored the coal operators and decided that the pollution of the streams by the waters pumped out of the mines is an evil for which there is no remedy for those residing in the coal area.

These communities are by nature admirably suited for the maintenance and propagation of a physically and intellectually strong people. Elevation, favorable atmospheric conditions, a moderate temperature and a plentiful supply of good water, offer a physical environment rarely found in any industry furnishing employment to an army of 140,000 employees. Reason would dictate that these favorable natural conditions should at least be preserved if not improved by man. Unfortunately, they have not been preserved, to say nothing of being improved. The foulness which surrounded the village among the kitchen middens could not have been greater than that which infests some sections of mining towns, where the streets are unclean and the rancid stench from decayed vegetation and stagnant pools poisons the air. Nothing proves the favorable natural conditions of these regions more than the fact that families flourish when surrounded by an atmosphere that is contaminated by their own indifference or ignorance. A primary law of hygiene teaches that an aggregation of men is liable to the action of no poison as virulent as that arising from its own neglect or ignorance. This law, so well established at great cost, is largely ignored by the average mining community. All classes of men are guilty, and a tribute of suffering and death is annually paid because of this disregard of the first law of well being.

These changes in the physical features of the coal fields illustrate how man can affect his environment, which, in turn, reacts upon him. Mining, carried on under the most favorable circumstances, is not advantageous to social progress, but if in addition to disagreeable work the face of nature is despoiled and the physical surroundings of homes are poisoned by carelessness, man must suffer both in mind and body. When men settled here a century ago, the life of nature was far more varied and wholesome than it is this day. Forest and stream

The Environment of a Shaft Showing the Refuse of Forty Years of Mining.

offered the huntsman and fisherman sport which strengthened his limbs and sharpened his wit. Many bright winged birds which are now rarely seen added to the charm of the landscape. The fish are exterminated, and the forests where solemn grandeur and majesty impressed the souls of men are no more. The children of mine employees are to-day raised among surroundings that are dismal and dreary. The hugh culm and rock heaps, polluted streams, bare and barren hills, cave-ins and strippings, make up the landscape which greets the eyes of these thousands, and if they are polluted in mind and body we need not be surprised. Man can turn a wilderness into a garden, but it needs intelligence and forethought. In these regions hardly a spot can be found in the villages and towns that is not cursed.

The blight is not equally cast upon every section of these coal fields. The Wyoming Valley, once called the garden of the State, still preserves some of its former beauty. Here many truck farms are found while many acres held by the coal operators are well cultivated fields. In the Middle coal fields, on the plateau on which Hazleton stands, there were once green fields which afforded ample pasture to cattle. Most of this is now torn by strippings, and the surface of this region to-day presents an amorphous appearance which baffles description. In the Southern coal fields, the surface is chiefly composed of sandstone rocks, where brush and stunted trees struggle to live. A portion of Schuylkill county is the most barren of all the anthracite coal area, and a few valleys, which formerly relieved the dreariness of the landscape, are now covered with coal dust.

Mining operations have depreciated the few farms which are still found in the Wyoming Valley. The rains, as above stated, pass through the soil to the cavities beneath. The vital forces of the soil are thus weakened and the agriculturalists find it necessary each year to spend a considerable sum in manure, in order to maintain the productivity of their farms. The soil of the Wyoming Valley is naturally fertile and, under ordinary conditions, yields a plentiful crop ; but the money necessarily spent by the farmers in recent years for purposes of fertilization places them at a considerable disadvantage in competing with their brethren outside the coal area.

The psychical and moral effects on society, arising from the mining industry together with the incidental defacement of the natural features of the environment as above described, cannot be statistically computed. We will in the course of our study find anomalies in the economic, social and political life of these communities, which will partly show the effect. Man, the center of all our civilization, is the chief concern of life. It is so in these coal fields. Adam Smith said: "Wherein consisted the happiness and perfection of man, considered not only as an individual, but as the member of a family, of a state, and of the great society of mankind, was the object which the ancient moral philosophy purposed to investigate." That same problem is to-day moving society as never before. Here, on these islands of anthracite, dwell 630,000 persons, forming a heterogeneous group which represents some twenty-six different races of men, and in each one of these the "divine shekinah" shines. To awake the man or woman in each of these units, is a task worthy of the best efforts of men. Natural scenery has been inevitably marred by the hands of men in the development of the anthracite coal industry, but to-day over 100,000 families find subsistence in a territory which formerly scantily provided the necessaries of life to a thousand, and where a hundred families of Indians could not live. It is the study of these aggregates that gives interest to these regions.

On this area, where representatives of twenty-six different races meet — all directly or indirectly depending upon mining, all subject to the influences of a society where customs and usages clash, all acted upon by a physical environment far from ideal — the question is what kind of an individual is produced? What is the type that is brought forth? After all is said, the individual must ever be the concern of society. Are the influences moulding the life of the descendants of immigrants to these coal fields in harmony with the great law of progressive social evolution? "In becoming more distinct from one another and from their environment, organisms acquire more marked individualities." This unquestionally is true of the descendants of European nations in the coal fields, but the point of vital importance is, are the "more marked individualities" resulting in an improved type? Notwithstanding many signs of retrogression, the law of social capillarity is pulling up the ignorant and impoverished immigrants of these coal fields, and it is safe to say that materially and intellectually their descendants occupy a higher plane in society than their ancestors did. Environment is an efficient factor in civilization, but ethnic qualities is a more potent one in the alembic of humanity. In these various nations there are latent energies which promise much as they blossom on soil where personal freedom, individual liberty and free education are the privileges of all. And we feel confident that these physical and psychical energies, which are called forth to action as never before, will enhance the economic worth of these men and enrich the country of their adoption.

Immigration and Social Capillarity.

In the last decade of the eighteenth century, the territory occupied by the anthracite coal fields had less than 8,000 population; to-day about 630,000 persons find subsistence here. Nitti says that on the barbaric stage one person needs 20 square miles of territory to furnish him subsistence, on the pastoral stage one square mile will support 2.5 persons, on the agricultural stage 100 persons can live on a square mile, on the industrial stage 400 persons can live on that extent of territory, and on the commercial stage a whole nation may thrive in a small country. The mining industry forms the economic basis of the population now inhabiting these coal fields. There are 370 persons to the square mile of the total area of the anthracite coal deposits, and 1,321 to the square mile of the coal basins. The supply of anthracite is limited, and the industry will pass through the three stages, growth, equilibrium and decline. During these several stages the number of persons capable of being sustained is limited, and as population passes or falls short of that limit, so will the economic life of this group be moderate or intense. The development of the coal fields, improvement in the art of mining, the introduction of ancillary industries, the increased wants of the people, have coöperated to augment the population up to its present number. There is a limit, however, beyond which the increase cannot pass. The Inspector in Chief of the Bureau of Mines said, in his last report, that the anthracite collieries have reached their maximum production. For some years to come the tonnage of these collieries will, under normal conditions, be 60,000,000 + tons annually, and about 150,000 mine employees will find subsistence therein. When the business declines, the surplus population will be forced out to other industries, but the process of expulsion will intensify the conflict within the mining industry, for the reason that the major part of the surplus labor seeking employment will not readily migrate. The ancillary industries depend upon the mines. Some factories locate in these coal fields because of cheap fuel. But these are of minor importance. They will not suffice to absorb the surplus labor population and relieve the tension that must inevitably come with the decline of the anthracite coal industry.

Anthracite mining is about 80 years old. This period has been one of growth. The population of the coal fields has also steadily increased during these years, though at a diminishing rate. The accompanying table shows the percentage rate of increase of population and of production at the mines, in the counties of Lackawanna, Luzerne and Schuylkill, in the decades from 1820 to 1900.

Year.	Population.	Percentage of Increase.	Production.	Percentage of Increase.
1830	48,123	53.6	500,000	400(?)
1840	73,059	51.8	1,000,000	100.0
1850	116,785	59.8	3,500,000	250.0
1860	179,754	53.9	8,500,000	143.0
1870	277,343	54.3	12,500,000	47.0
1880	352,308	27.0	24,800,000	98.4
1890	497,454	41.2	40,000,000	61.3
1900	623,879	25.4	54,000,000	35.0

The percentages of both population and production show a decline. The decade, 1870–1880, was one of great industrial friction, caused chiefly by the readjustment of wages after the inflation of the currency during the sixties. In the last decade, the percentage of increase in population and production is lower than in any previous one. And without presumption, we may safely predict that the percentage increase will be still smaller in both columns in 1910.

In the last decade the rate of increase of population was not uniform in the three counties chiefly dependent on mining. While Lackawanna increased 36.4 per cent. Schuylkill only increased 12.2 per cent. In the latter county, mining is practically the only industry which furnishes employment to the people, while in the former the city of Scranton enjoys a variety of industries, which favors an increase of population not found elsewhere in the anthracite coal fields. Of the 32,813 mine employees in Lackawanna county, only 8,485 are in the city of Scranton, while another army of over 24,000 males find employment in that city in the various professions, trades and commerce, manufacturing and mechanism, etc. But in Schuylkill county, out of 49,592 males engaged in occupations, 33,228 are engaged in mining, while the remaining 16,364 are engaged in various pursuits supplying the wants of the mining population. In Lackawanna county, 53.23 per cent. of the male workers are mine employees; in Schuylkill, 67 per cent. are so employed. So that the latter county gives us a truer percentage of increase in the mining population during the last decade than the general percentage for the three counties.

Immigration into the anthracite coal fields has virtually ceased. A large percentage of the increase from 1880 to 1890 was due to Sclav immigrants. These peoples furnished the cheap labor needed in the development of the thinner veins of coal, and supplied the operators with men willing to work under conditions which labor of a higher grade resented. The Sclavs represent possibly the lowest grade of European workmen that can be imported, and Congress being in its present frame of mind, it is not likely that oriental countries will supply operators with labor still cheaper, which would displace the Sclav as he displaced the Anglo-Saxon, the Celt and German.

The increase of population in the anthracite coal fields in future will depend chiefly upon natality. The present population is amply sufficient to furnish the necessary labor for the maximum tonnage the collieries can produce. Conditions in the industry will not improve, and for many years past they have ceased to attract British miners. The law of social capillarity will continue to raise the social standard of descendants of foreign born parents, and a certain percentage of them will leave the mines, but the birth rate of the Sclav population will more than supply the labor needed in an industry which will soon be declining. For these reasons, we cannot expect future immigrants to our country to settle in the anthracite coal fields to any great extent.

By the last report of the Bureau of Mines, there are 143,826 mine employees. This number of males earning wages gives us a population of about 450,000 directly dependent for a living upon the mines. Another army of about 60,000 male and 20,000 female wage earners are engaged in the professions, trades, transportation, domestic work, manufacturing and mechanism. These avenues of employment are mainly ancillary to the chief industry of the anthracite area, and sustain a population of about 180,000 indirectly dependent upon mining. Thus within the anthracite coal area we have a total of 630,000 persons, mediately or immediately deriving their subsistence from the production of the anthracite collieries.

In the cities of Scranton and Wilkesbarre, females form 22.7 per cent. and 23.6 per cent. respectively of the wage earners. Outside these two cities, in the area under consideration, the females earning wages to aid the family in the conflict of life do not exceed 5.5 per cent. of the total wage class. That is due to the want of manufacturing industries, which are, comparatively speaking, few in number outside the cities mentioned. Hence the mining industry is peculiar in this regard, that it casts the burden of supporting the family upon the male members. This fact necessarily influences the economic and social conditions of the youth of both sexes in mining towns. The boys are put to work at an early age in order to increase the family income; the girls, wholly dependent on others for their maintenance and having no way of earning an independent competency to supply their wants, are on the one hand subject to serious temptations because of lack of means, and on the other tend to enter inconsiderately into marriage to relieve the economic tension of the home. It also places the average family in mining towns as compared with that in cities at a disadvantage, because of the want of industries wherein the females may earn a competency. Families in mining towns find it easier to carry on the conflict of life when the children are male than when they are female. Boys can find employment, but few are the avenues open to girls in which they can make a living. In every town and village there is a supply of energy which could be productively used, if certain lines of in dustry could be introduced. As things are, the energy of nearly half the youths of our area is not productively used, and the economic, social and moral progress of these groups will not be secured until they are wisely directed into channels of usefulness.

The foreign born immigrants and the majority of their descendants, as above stated, depend upon the production of these mines for their subsistence. Few are the industries which could be carried on in these communities if coal were not conveniently and cheaply furnished them. The lean and fat years in these regions can be estimated according to the amount of coal shipped to market in exchange for which our people secure the necessaries and enjoyments of life. During the last thirty years the number of tons produced per mine employee varied from 252 to 405. The average production per mine employee, from 1870 to 1901 inclusive, was 343 tons per year. The part of this productive wealth given to laborers maintains them and their families. The adult laborer's share does not, on an average, amount to $500 a year. From this sum the necessaries of life must be procured, a family raised, and whatever comforts and enjoyments mine employees have these must also come from the same source.

Much has been said of late as to the inadequacy of this sum to enable the laborer to secure the American standard of living;

little is said of the productive efficiency necessary to attain that end. To attempt to fix the laborer's income by a standard of living, regardless of his productive power, is to attempt the impossible. The progress of society depends upon increasing the productive capacity of the individual so that he will be able to earn more than bare subsistence for himself and family. And the fact that a large percentage of mine employees have been able to earn wages in this industry amply sufficient to raise the social standard of themselves and family, is a proof that a capable workman, by thrift and industry, can do more than simply conserve the bodily existence of himself and those dependent upon him.

The aggregate wealth annually produced in the anthracite coal mines has made this area one of the best business sections in our State. The commercial prosperity of our communities will undoubtedly decline as the sum total of wealth from the collieries diminishes. The following table gives the approximate number of tons per capita of population dependent upon anthracite mining produced at the collieries for the years mentioned.

Year.	Tons per Capita.	Year.	Tons per Capita.
1840	13.7	1880	70.4
1850	29.8	1890	86.3
1860	47.3	1900	86.4
1870	45.1		

Up to 1890, with the exception of 1870, which was an abnormal year because of industrial friction, the per capita production of the mines shows a vigorous growth. The last decade stands in striking contrast with this. Here a stage of equilibrium is reached, and the next decade will see a reduction in the number of tons per capita of population directly or indirectly dependent on the anthracite coal industry.

This fact will have an important bearing upon the financial and commercial life of these communities. A financier, who has been in business for many years in the coal fields said, "Business here is not what it used to be." When asked what was the reason for the change, he replied: "In former years the operators lived in the coal fields and spent their money here, but of late they are absentees." This may account for a fractional part of the depression, but the prime source lies in the fact that the collieries having passed through a period of extensive evolution and are now in the intensive stage, and the aggregate wealth produced, estimated in per capita of the population, is not so large as it was in former years. From 1870 to 1895 the building and loan associations did a thriving business in the coal fields. In the last decade their business has fallen off considerably, for many of our towns have not increased in population for the last six years. A large birth-rate will continue to add to the population, and as the necessaries of life will make an ever-increasing demand upon the sum total of wealth produced, business men who handle goods of positive utility will find increased trade, while those who handle goods of neutral utility will be confronted by market conditions which grow annually more stringent.

The conflict for subsistence will also grow more intense to the individual and the family in the coal fields under the pressure of the Malthusian law. The tension will undoubtedly be relieved by the migration of the young and the enterprising to other parts of the country. Labor moves freely among us, but the ties of home and family, social affiliations and pecuniary interests will so retard the movement that the intensity of the conflict will be felt by our people and will become the occasion of friction in the industrial and political life of these communities.

The elements of our population have been, during the last thirty years, in a condition of flux, which is not advantageous to social progress. Mining is disagreeable work. The thrifty members of our society rise to more congenial employments. Hence, annually, there is a withdrawal and migration of the élite of the group, and the loss thus caused to mining communities is intensified by the influx of less advanced elements as well as by the natural operation of the Malthusian law. To secure the progressive advancement of society in the anthracite coal fields under these conditions is very difficult. As soon as the receptive members of mining towns are trained in and acquire the principles upon which the social fabric of our institutions rests, they withdraw to the cities or migrate, leaving the work of schooling the less apt as well as the new individuals added to these communities to the few who are moved by patriotic or philanthropic motives to undertake it.

The character of the population of this area has perceptibly deteriorated in the last thirty years. A selection has been effected but in a retrogressive sense. The physical strength of the accretions of the last quarter of a century may favorably compare with that of any previous period, but their intellectual and moral qualities are decidedly lower. This has not resulted wholly from the influx of needy, ignorant and incapable workmen, but also from the degeneracy of the descendants of the shiftless and thriftless immigrants of past generations. These, too lazy to work and too poor to emigrate, lead a parasitic life and give nothing in return to the communities but corruption in politics, agitation in industry, and debauchery in society.

In every healthful community there are two elements, viz., the conservative, which opposes change and is anxious to keep things as they were, and the liberal which is anxious to change things according as the environment changes. When these elements are kept in approximate equilibrium, so that vigorous conflict ensues, the organism is strengthened and progress follows. If either of them so triumphs that conflict ceases then disorganization and degeneracy are the consequences. Now, the withdrawal of the élite members from anthracite communities means the weakening of the conservative factor, and the readier subjection of those remaining to the crudities and vagaries of the ignorant and the designing. Well has it been said that "the chief errors of the world, as well as its chief evils, have a common origin in ignorance." Many of our communities greatly suffer because of the absence of a strong, intelligent, honest and conservative element. The so-called liberal factor in mining towns abuses power, ignores precedents and is ignorant of the principles of social health.

It has no regard for facts, no veneration for learning, no gratitude for benefits and no consideration for the historical foundations of the municipality and the state. Unconscious of its own ignorance and incapacity, it substitutes blatant arrogance for efficiency and wilful daring for careful and competent investigation. This lowest stratum of the liberal factor disturbs the peace of these communities, leads to culpable wastefulness in management of public finance, and confounds industrial relations so that the mining business is either raised to the verge of a revolution or debased to the condition of abject servitude.

The Twenty-six Peoples Now Residing Here.

The labor required for the development of the anthracite coal fields has been furnished by immigrants. In the first fifty years of the development of the industry, the British Isles and Germany furnished the supply. In the last twenty-five years the Sclav nations of southern Europe have done so. The change in the character of the immigrants to the United States is synonymous with that which has taken place in these coal fields.

The following table brings out very clearly the increase percentage of Sclav and the decrease of British and German immigrants to the United States, as given by the Commissioner of Immigration.

Year.	Sclav and Italian.	British and German.
1861–70	1.05%	77.38%
1871–80	6.44	57.46
1881–90	17.65	52.72
1901	68.50	13.50

In the census of 1870, the Sclav element in Lackawanna, Luzerne and Schuylkill counties was hardly perceptible. In the last census they numbered 72,748.

In 1880, not 5 per cent. of the mine employees were Sclavs, to-day about 50 per cent. of them belong to these races. In the report of the Bureau of Mines for the year 1900, returns from 232 collieries showed 42.31 per cent. of the employees classified as "non-English speaking peoples." In this computation the Sclav breaker-boys, most of whom are native born, are not classified as "non-English speaking." These, however, are Sclavs in training and sentiment. The mine inspector of the Fifth District had, in 1902, 51.24 per cent. Sclavs among the employees in his territory. We can safely say that 50 per cent. of the miners and laborers in the anthracite collieries are Sclavs, which gives us between 34,000 and 35,000 adult males of these nations. Half of these are married, and being from necessity endogamous, their wives are also Sclavs. The 17,000 Sclav families give us a population of 85,000, and the 17,000 bachelors added gives us in the anthracite coal fields over 100,000 of these peoples.

This change in the character of the labor force of the anthracite coal fields well illustrates a Gresham law operating in the labor as well as in the financial world. The Anglo-Saxon * mine employees, in the early eighties of the last century, felt the operation of the law and tried in various ways to ward off its effect. Laws were passed to create boards of examiners to issue certificates of competency to miners. An apprenticeship of two years is required before a laborer can become a miner, and many think that familiarity with the English language is a necessary qualification to mine coal. Social barriers also have been erected against the Sclav, but all to no effect. A silent but steady exodus of the most intelligent and capable mine workers goes on, and simultaneous with it, the Sclav, conscious of his inferiority, adapts himself to the conditions and thrives.

The immigration of the Sclavs into the anthracite coal fields is well illustrated by the following table, relative to their settlement in the cities of Scranton and Wilkesbarre.

* The word Anglo-Saxon is used in this work to designate the English-speaking mine employees, and the word Sclav to designate those who are generally called "foreigners" and have little or no command of the English language. This use of the word, though not ethnologically correct, gives us a simple means of contrasting these two elements of our population without naming the various races in the groups.

SCRANTON.

Year.	Austro-Hungary.	Poles.	Italians.	Russians.	Bohemians.	Total.	Percentage of Foreign Population.
1870	7	15	7	0	0	29	.18
1880	28	67	12	27	4	138	.87
1890	1,106	600	367	488	11	2,572	9.00
1900	1,390	3,750	1,312	671	63	7,186	24.80
WILKESBARRE.							
1890	382	393	23	149	29	976	9.57
1900	601	1,632	189	469	19	2,910	23.85

The steady flow of Sclavs and Italians into the population of these cities is synonymous with what has been going on in the anthracite coal area. The change has been more thorough, however, in mining villages and towns than in the above cities. Many mining camps, which in the seventies were inhabited by Irish, English and Welsh, have passed wholly into the hands of the Sclavs, while every mining town has within it a colony wholly composed of "foreigners." In 1870, of the 38,161 foreign born persons engaged in mining in the State of Pennsylvania only 121 or 3 per cent. were Sclavs and Italians. By 1890 the proportion in the anthracite coal fields was 25.67 per cent. and in 1900 it had reached 46.36 per cent.

The foreign born peoples forming about 32 per cent. of the total population of our area represent 26 different nationalities. They are English, Welsh, Scotch, Irish, German, Swedes, French, Swiss, Dutch, Poles, Sclavonians, Austrians, Hungarians, Bohemians, Tyrolese, Russians, Lithuanians, Greeks, Italians, Hebrews, Negroes, Arabians, Cubans, Mexicans, Spaniards and Chinese. The last seven mentioned form an insignificant portion of the total population. The Sclavs and Italians would form about 15 per cent., the Anglo-Saxons and Germans 17 per cent., and the remainder 68 per cent. native born. If, however, we classify the native born of foreign parentage with the foreign born we have over 70 per cent. of the population in that class.

Some economists have said, that the labor necessary for the development of our industries would have been supplied by the natural increase of the colonists if immigration had not taken place. This statement loses sight of the operation of the law of social capillarity and the unpleasant nature of some work necessarily connected with industrial development. The work of mining coal is dangerous and unpleasant, and descendants of miners as a rule in our country cherish a strong aversion to working underground. The young men, who graduate from the public schools, will not enter the mines if they can possibly earn a living elsewhere. Those of this class, who gain a subsistence in the collieries, perform the lighter work. Mine foremen invariably say that the native born mine employee shirks hard work. Dr. A. G. Keller says that the experience of the German colonists is that a highly educated negro is a "Schurke" and absolutely useless for all practical purposes. The consensus of opinion among superintendents and foremen in the anthracite coal industry is that the mines could never be operated if they depended upon the native born for the labor supply. It would be well for the censors of foreigners to remember this. These immigrants are the hewers of wood and the carriers of water in the land, and entrepreneurs would find it far more difficult to carry on their operations if the supply of foreign labor were not at hand to perform menial toil which is shirked by wage-earners on a higher social plane.

The labor, which has developed the anthracite coal fields has virtually been wholly supplied by emmigrants and their descendants of the first generation. This class in our area form over 70 per cent. of the total population, while in the country at large they form only 32.93 per cent. In the census of 1874, of the 152,107 persons engaged in mining, 94,719 or 68.80 per cent. were foreign born. If we add to the foreign born the number of native born of foreign parentage engaged in mining, the percentage will be much higher.

Year.	Foreign Born.	Per cent. of Population.
1860	49,753	27.70
1870	85,544	30.84
1880	88,779	25.20
1890	142,035	28.55
1900	161,357	25.86

The preceding table gives the number of foreign born in the counties of Lackawanna, Luzerne and Schuylkill for the years specified.

This gives an average of 27.62 per cent. of the total population of the above counties as foreign born. In 1870 in the three mentioned counties, when the foreign born were of Saxon and Teutonic extraction, 56.42 per cent. of the total population was either foreign born or native born of foreign parentage. In 1900, when nearly half those of alien births are Sclavs, we have 63.13 per cent. of the total population either foreign born or native born of foreign parentage. An investigation into the nature of the population of thirteen purely mining towns located in the Northern coal fields, resulted in 32.77 per cent. of the total population being foreign born, and 72.22 per cent. either foreign born or native born of foreign parentage. It is safe to say that an average of 70 per cent. of the 630,000 people in the anthracite coal fields is either foreign born or native born of foreign parentage. In other words, 441,000 of the total population in the area under consideration are either foreign born or native born of foreign parentage. These have furnished the labor necessary to produce the annual tonnage of anthracite coal sent to market. There is much that is socially and morally undesirable in the foreigners engaged in mining, but the fact that the brawn and muscle necessary for the production of coal has been furnished by them should never be lost sight of.

It is not long since when the mines of the civilized world were manned by serfs, slaves and convicts. The last class of serfs on British soil to receive emancipation from conditions which made them little better than chattels was mine employees. This possibly accounts for the public sentiment which places a low estimate on mine workers. In the hierarchy of labor tradition has assigned the "colliers" a place low down in the social scale, and public opinion in a country where manual labor is more highly honored than in any other land still clings to that sentiment. This is due to bias and ignorance of mine employees. They deserve greater honor as a body of able men in the army of producers. Few classes of workers sacrifice as many lives as they do in bringing their portion to the national fund, and, considering the nature of their calling and the danger incident to it, this class of workers does not take an undue proportion of the store of wealth produced by its labor.

This heterogeneous confluence of so many European nations has a marked influence upon the economic, social and moral life of these communities. L. F. Ward has said that "the condition of the European race is such now that in point of average capacity there is probably, except in isolated localities, no distinction in the different ranks or social stations in life." From a scientific standpoint few would take exception to this statement, but in the practical affairs of life, where racial pride and prejudice play so great a part, few indeed would concur with this view. Nothing is so conspicuous in communities where the races of men mix as the ethnic confidence found in each group. The social standing of many shiftless Anglo-Saxons in these communities is the lowest imaginable, and yet they ever insist upon their superiority to the Sclav and are indignant if classified with him. In a miserable mining patch we found an isolated English-speaking family among many Sclavs. The conversation had hardly passed beyond the usual exchange of courtesies when the woman pointed with contempt to her neighbors and said: "We don't have nothing to do with them." Estimating the social status of the speaker by her personal appearance and the surroundings of the house in which she lived, the lowest Sclav would not be improved if she did associate with her. This ethnic pride is also found among the Sclav. The impoverished Magyar always insists upon his social superiority to the equally impoverished Hun. The Pole looks with contempt upon the Lithuanian and the latter is prompt to assert his claim to a more remote ancestry and an older civilization than the former. This racial pride, equally strong in each race, is the cause of many conflicts between these men when they meet over their cups. In the early years of mining it precipitated many a conflict between the immigrants of the various races from the British Isles, and the bloody and

Sclavs in a New Environment.

fatal quarrels, which so frequently take place among the Sclavs are due to the same cause. It has its influence upon the industry. The Pole and Lithuanian will not work together. Foremen have to study national proclivities and prejudices with regard to the productive efficiency of groups of employees under their management. In large towns, where the mine employees live, the various races form colonies and generally keep within the limits of the section appropriated by them. Hence we have "Scotch Road," "Murphy's Patch," "Welsh Hill," "Hun Town," "Little Germany," "Little Italy," etc. In these various sections national customs and usages are perpetuated, and ethnic peculiarities do not cease when we enter the moral and religious spheres.

The physical endurance of the average Sclav is unquestionably equal to that of the average workman of any nation. His intellectual capacity clogged "by the weight of centuries" is not equal to that of the German and British workman. In speaking of the Russians, Mr. Ward said that they are a few centuries behind the rest of the civilized world, but that "there is no people on the globe more capable of making proper use of knowledge if presented to them." This is true of our Sclav population, and we can undoubtedly look forward to a rapid development of the physical and intellectual energies of these peoples, under the benignant influence of an environment which favors social progress.

It is impossible for these representatives of various European races to meet in industrial and social life, no matter how great the barrier of national pride, prejudice and suspicion be, without each unconsciously influencing the other. And the ac ion of these forces of competition and conflict will undoubtedly awaken the energies of the Sclav to greater efflorescence than they ever manifested in the fatherland. Unfortunately for the Sclav, as well as for society, the element of our population most accessible to him is not the best. A certain percentage of the early immigrants has not improved its opportunities. It has degenerated. Improved economic conditions only furnish them with larger means to gratify appetite, and confirm them in vice and sensuousness. These men form the dregs of society and drift into the most disreputable sections of our towns. The Sclav, partly because of his low standard of living and partly because of social ostracism, is also driven to dwell in these sections of mining towns, and if he is inclined to imitate the English-speaking peoples, the type nearest him is vile and uncomely. The corrupting influence of this small and degenerated group of Anglo-Saxons is far greater than its numerical importance would warrant. The better class of citizens stands aloof, and the attempt of aspiring members of the Sclav races to move into the better sections of the towns is speedily resented. "It depreciates property," the good people say. They are blissfully unconscious of the human depreciation that goes on in vile quarters where the bestial passions of Anglo-Saxons defy the claims of decency and propriety, and designing men lead a parasitic life trading upon the ignorance and weaknesses of the Sclavs, far removed from the wholesome influences of home and the restraints of parents. If the Sclav is to be improved the conservative and intellectual class in the mining towns ought to exercise greater consideration for his aspirations, and the deadly influence of the parasites now feeding upon his weaknesses should be counteracted.

Walter Bagehot said: "In the early world many mixtures must have wrought many ruins, they must have destroyed what they could not replace — an inbred principle of discipline and order." The mixtures in the anthracite coal fields cannot be studied but the conviction grows that the moral restraint due to discipline and order is largely wanting. In every town differences on the main points of human life are tolerated. It cannot be otherwise in a society where customs differ. But this tolerance results in carelessness of manners and leads to skepticism. Many are led to believe, as they see peoples diverge on moral questions, that right and wrong are the creatures of human caprice, having no objective validity in the well being of society. The native born questions the traditional ideas imported from beyond the seas, which were potent factors in the lives of his parents and tended to the preservation of law and order. The inevitable consequence is moral retrogression. It is patent to all and leads to disastrous consequences in the lives of native born of foreign born parents.

The great need of the hour in mining communities is the introduction of intellectual and moral forces which will counteract the tendency to retrogression from this mixture of races. The United Mine Workers' Union is a beneficent and potent factor in obliterating racial suspicion and prejudice. If the organization is preserved and its usefulness enhanced, it will, more powerfully than aught else, lead the way to social progress and assimilation. The Sclav and Anglo-Saxon understand each other better when they meet in discussion over questions of economic amelioration and coöperate for improved industrial conditions. The intellectual and moral interests of the mine employees are not furthered by the above organization as is their economic. If they were, we would be more confident of the progressive and permanent improvement of society as found in these coal fields. Lilienfeld has said that the progress of society can only be secured when a simultaneous advancement is made in the economic, political and juridical spheres of human activity. The improvement in the economic life of these peoples is unquestionable, but the anomalies in the juridical and political life of anthracite communities are ominous, and unless arrested they will work the ruin of many.

The native born children of foreign born parents furnish good material to work upon. They know little of the prejudices and antipathies of their parents, and promiscuously playing and wrestling on the streets, learning the same lessons and singing the same songs in school, they grow up a homogeneous people eager to serve their country and quick to grasp the advantages offered for social advancement. Under the magic touch of American civilization we can confidently hope for a physically and mentally improved type among the descendants of the Sclavs. All we need is to secure free opportunity for American institutions and ideas to play upon these receptive minds and hearts, and no children raised in our public schools will make better use of them and yield better results.

CHAPTER II.

THE SCLAV EMPLOYEES.

The Jews who Accompany the Exodus.

Wherever the Sclavs have settled in these mining towns a complement of Russian and Polish Jews is invariably found. The returns of the Commissioner of Immigration show that these immigrants are the poorest of the poor invading our land, but no sooner are they in these communities than they begin to accumulate riches and some of them to-day are among the wealthiest in our mining towns. The Jew's social status is as low as that of the Sclav and many of them adhere to the standard of living which they brought with them from the land of their birth, notwithstanding their material prosperity is greatly increased. They raise large families, are tender fathers and faithful husbands. The males are rugged and well developed; the females strong and the mothers of children.

The Jew is seldom found engaged in manual labor in and around the mines. He settles among the Sclavs, lives as they live, is as filthy as any of the Hungarians, but engages in business or carries his pack from door to door. He knows the Sclav, can converse with him in the various dialects, can cringe as low as any of them when soliciting business, can cater to his appetite, and knows exactly the taste of his patron. A few of them practice trades such as shoemaking, tailoring and plumbing, but most of them are merchants of some kind or other. The poorest carry a pack of dry-goods from house to house. If they prosper they buy a horse and wagon and thus supply a larger number of patrons. Then they open a store in one room of the home they own and in the course of years are found on the main street in possession of a flourishing business. Thrift, industry and perseverance are their characteristics, and their keen business sense brings them success when tradesmen and merchants of other nationalities are unable to live. The clothing trade is almost wholly in their hands and the dry-goods business is rapidly being monopolized by them in our territory. Many of them engage in the wholesale liquor business, sell to families in small quantities and are not particular to keep within the letter of the law. We have known some of these men engaged in private banking for from ten to fifteen years, to have handled thousands of dollars annually, not a cent of which they ever lost and never were they guilty of a breach of trust with the Sclavs who confided in them. One of the most prosperous of these sent on an average $60,000 annually in the last decade to these people's ancestral homes and in the year 1901 he had from $20,000 to $25,000 of their money in his possession. He rarely paid interest unless the depositor had to his credit several hundreds of dollars; then he paid two per cent. The men could get their money at any hour, and often after a fight on pay-night, the private banker was called up at midnight to cash the account of a fugitive. In the town of Olyphant, of 13 heads of Jewish families, only two were without real estate; the others were worth from $2,000 to $50,000. The same is true of this people in other towns in these coal fields.

The Polish and Russian Jew belong to the orthodox branch of the Jewish faith, and carry out with minute exactness the ritual of the orthodox Hebrew. The skema is recited daily, the philacteries are just as regularly bound around the arms and forehead, the thorah is faithfully taught the children and the studies pursued in the public schools are not allowed to interfere with this parental duty. Wherever from ten to fifteen families live, there a synagogue is erected, a teacher engaged and the children daily drilled by him in the faith of their fathers. All the Jewish children attend the public schools, but in addition to this they spend from two to three hours in the private school under the care of their religious teacher. The orthodox Jew will not work on Saturday, but as soon as the sun is set he is ready for business. The trade they do with Sclavs on Sunday, after mass, is very great, and it is a matter of general observation in these communities to see Sclavs carrying bundles — wet or dry — as they return from their devotions. A Hebrew, engaged in the wholesale liquor business and a gross violator of the license laws of the State, when questioned as to the religious life of his people, showed us his arm where the philactery had been bound that day and said "Yes, we Jews are religious." One of these private bankers, doing business with the Sclavs, kept on tap a quarter barrel of whiskey which was an admirable lubricant in the transaction of business. The guileless Sclav is often ready material to the hand of the Jew. One of these men, about to return to the fatherland, drew $500 from a private banker on which he had received no interest and also bought his railroad and steamship tickets from him. The Jew gave his daughter, twelve years of age, a present of a pair of shoes, for which the father was very thankful. The Jew, however, knew his business, and when asked about the transaction by a friend, said, "Yes, I was good to him. I gave the girl a pair of shoes and charged him extra for the passes."

The safe-keeping of funds for the Sclavs often means a great deal. Many of them in mining camps and towns incur great anxiety and trouble to safeguard their money. Some hide it in the mines, others sew it in their garments, others hide it in their trunks or secrete it in the bedding, etc. In all these cases it is a source of constant anxiety and they know not when it may be stolen. Under such circumstances the reliable private banker serves a useful purpose and the Sclav is willing to waive the demand for interest, if only he is confident of the safe-keeping of his money, and can get his savings when he needs them.

Many of these Jews also make their linguistic capacity profitable. In the frequent broils among Sclavs an interpreter is necessary in the local court, and the Hebrew is called upon. And not only are his services necessary as an interpreter, but also he is in demand as bail. His compensation as interpreter is fixed by court, but his charges as bail are often outrageous. Some of these Jews are influential enough to stand between the indicted and the prison cell, but for the service the criminal pays a fancy price.

As a class the Jews are thrifty and industrious, peaceful and temperate. In the liquor business they are never incapable of doing business. If a barrel is kept on tap the contents has no temptation to them — it is kept solely for business. They are mercenary and sometimes violate the law by embezzling, or receiving stolen goods, and yet, generally speaking, the Jew is seldom implicated in crime. They occasionally quarrel among themselves and in the heat of conflict enter a law suit, but they invariably settle the dispute and save the costs of prosecution. The Hebrew is conservative and his love of riches and anxiety to acquire wealth keep him longer at a low standard of living than the representatives of other nationalities. He is more clanish than the Sclav and is less susceptible to the influences which mould and elevate the lower races which daily mingle with Anglo-Saxons in industrial and social life. The Hebrews form a colony within a colony, and only come in contact with other nationalities in business. When their sphere of business is among the Sclavs, it is easily seen how they are slower to respond to the demands of civilization than nations with less business capacity than they.

The Sclav's Home in the Fatherland.

The Sclavs of Europe occupy the eastern portion of the continent which is bounded on the east by the Baltic Sea and the Ural Mountains, on the south by the Black Sea, on the north by the White Sea, and on the west by the German Empire. They are divided into two groups, the northern and the southern, both of which are brachycephalic, but the southern differs from the northern in being taller and darker. Those who are in the anthracite coal fields come from the regions bordering the Baltic Sea on the north and extending to the south as far as the Black Sea, including the province of Galicia which is the southwest section of the Russian Empire and Austria. The nations represented are the Letts, Slovaks, Ruthenians, Hungarians, Magyars, Poles and Bohemians. The Slovaks, Hungarians, Magyars and Poles come from provinces subject to Russia, Austro-Hungary and Germany. They are broad-headed, their complexion light, their body compactly put together, their shoulders broad, their chest deep, with well developed arms and legs. Many of the Poles are brunette which suggests a mixture with the dark races of the Mediterranean. The Little Russians or Ruthenians come chiefly from Galicia and are subject to the Czar. Their dialect differs considerably from the Russian language, but they belong to the Russian family and have furnished the Czar the Cossack troops which have played so important a part in European history. The Ruthenians clearly show a mixture of blood, for many of them are darker and smaller than the typical Sclav, while the shape of their skull and their psychical characteristics differentiate them from the Slovak and the Huns. The Letts cannot be exactly classified with the Sclavs. They are said to be a part of the northern branch of the European Aryans and are considered to be among the early inhabitants of Europe prior to the invasion of the Sclav. Their leaders claim their people to be one of the aboriginal races of Europe, which settled in that country previous to the immigration of either Teuton or Sclav from the steppes of Asia. The nation has a romantic history and, having stood for centuries the onslaught of successive bands of Teutons and Sclavs in quest of a better home than the steppes of Asia, they were forced to occupy some of the most inhospitable regions on the continent of Europe. When the Letts lost their independence, the Russian government compelled them to learn the Russian language, nevertheless they have preserved their native tongue which some linguists believe to be one of the most archaic members of the great Aryan family. The male members of the race can speak both Russian and Lithuanian, but the female as a rule have command only of the latter. For many centuries the Letts and the Sclavs have freely mixed so that they are to-day classified as one. The former are blonde and some of them are as fair as the Swede. The men are tall and make a more favorable impression than the Hun or Slovak. Their standard of living is as low as any, while the brutal fights, which frequently occur among them, confirm the opinion that they are as savage as any class in the coal fields. We have also a few families of Bohemians which clearly show the Sclav type, to which they are bound by both social and traditional ties.

We have many Italians in our territory who are classified with the Sclav as cheap labor. The Italian is very rarely found working under ground. Many of this race are found in the Fifth Mining District and have nearly monopolized the work in stripping mining. They also work on railroads, in works of excavation, while the trade of shoemaking and repairing as well as the conducting of fruit stands is rapidly passing into their hands. We have many Tyrolese also, who work underground and are expert rockmen. They are not so clearly distinguished from the Sclav as the Italian and are generally regarded by our people as belonging to the Sclavs.

Nearly all these peoples have come to the coal fields from agricultural communities. Some of them have been raised under the allotment system which prevailed in feudal times. They are all familiar with agricultural labor which is manifest in the interest they take in gardening, and, during the strike of 1902, many of them willingly migrated to farms where they were occupied as common laborers. They were raised on coarse fare on the farms of their native country and from their youth were inured to hard work and hard living. Their principal articles of diet were rye or barley bread, oats, beets, cabbage and potatoes. Meat was a luxury, and most of them considered themselves fortunate if they got it once a week, which generally fell on Sundays. Cattle and sheep were raised on the farms but these were generally sold to get money to pay the rent, the taxes, and supply the family with articles of clothing which they themselves could not produce on the farm. The clothing they wore was home-spun. The raw material was gathered in the summer, prepared into thread in the fall, and woven into cloth in the winter months. Adults tried to get one new suit each year.

Practically all our immigrants belong to the laboring class who lived in houses provided by the landlords. Attached to the home in the fatherland are four or five acres of land which are cultivated by the tenant and form part of his wages. Very little money circulates among them and they are generally ignorant of the use of money. The head of a family, living on the estate and having the use of a plot of ground, hires himself out for $25 or $30 a year. A young man works on the farm for a gulden a day, which is about 41 cents in our money, while in summer he works from sunrise to sundown for 50 cents. Women there dig, wheel, carry burdens and do general farm work for 12 cents a day, which in summer time might rise to 16 cents. Of course living is cheap. One can buy a pair of hand-made shoes for two or three guldens, while the best boots are sold for about a $1.50 a pair. Good board can be got for 10 or 12 guldens a month, while most families raise sufficient vegetables for their use. Under such conditions the struggle to live resulted only in bare subsistence to the wage-earning class, and the energetic soon availed themselves of the promise of better things in a distant land.

To these men nine hours a day in the coal fields of Pennsylvania at $1.50 or $2.00 a day is a great inducement. The Sclav standard of living in these coal fields is low, but if compared with that in vogue in his home, it is rich and varied. Bread made of the best wheat flour the market affords, meat of some kind every day unless religious considerations forbid it; on fast days eggs, sardines and cheese take the place of meat; a plentiful supply of cabbage, potatoes, pickels, apples and coffee —these give the Sclav a sumptuous living, while they drink beer more freely than they did water in the hills of their ancestral home. It is not astonishing that many of these people regard this land as a "goodly country" where they have realized more of the good things of life than was ever dreamt of in the fatherland, and the money and the letter sent to the folks across the sea, stating the wages and the living enjoyed by the immigrants in these coal fields have been the chief instruments to induce the enterprising Sclavs to come here and seek their fortune.

These economic considerations which have lured the Sclavs into these coal fields have also in recent years, effected a complete change in their plans for the future. When first they came, their ambition was to save enough money to buy a farm in their native home, return to the fatherland, and live in peace under the shadow of their own vine for the remainder of their days. Many did this, but it did not take long for the practical Sclav to see that if this was a good land for his material interests, it was equally so for those of his children. No sooner was this conviction reached than he acted accordingly. The volume of money sent to the fatherland perceptibly diminished, and they began to buy lots and build homes with the intention of making this country their permanent home. This change of sentiment has gone on for the last ten years and the Sclavs to-day own thousands of houses in the anthracite coal fields. The statistics of marriages given in a subsequent chapter show to what extent this change goes on. The spiritual leaders of these people also advocate this policy and the energy displayed by their parishioners in following it proves that the Sclav has come to stay. They buy real estate and raise large families, which will enjoy larger opportunities and higher privileges than could be realized in countries less advanced in civilization.

Do the Sclavs Make Good Miners?

In the chapter on population we saw that there are to-day about 100,000 Sclavs and their descendants in the anthracite coal fields. The Poles are the most numerous and were the first settlers in these regions. As a factor in the operation of these collieries the Sclav is indispensable. His political importance is daily increasing, and if aided by means whereby his social worth may be enhanced, he is capable of taking his place in the ranks of more highly civilized immigrants to our country. They will soon be masters in the political management of our towns, leaders in the commercial life of our communities and a determining factor in the industry upon which we all depend. This part of our population, upon whose character depends to an ever-increasing degree the peace and progress of our communities, is a promising field for the labors of patriotic and public-spirited men, whose duty it is so to replenish their minds that they may the better resist the vagaries of socialists who find in the foreign elements in our country the readiest material for socialism.

The Sclav is a good machine in the hands of competent directors. He is obedient and amenable to discipline, courageous and willing to work, prodigal of his physical strength and capable of great physical endurance. He is a good friend and an implacable foe. He thinks slowly and is willing to follow the lead of others, but when the Sclav is once set in motion in a given course, he is there to stay. His confidence in competent leadership is absolute, and both in work and in society he is quick to copy others. Sclavs are far better mimics than their English-speaking neighbors, and their children in school excel Anglo-Saxon children in penmanship, drawing, mathematics and discipline. They are fatalistic to an astonishing degree, which accounts for their stoicism in suffering or calamity. Gross superstitions are found among them, while their loyalty to the Church is more fanatical than intelligent. Religion to many of them is wholly severed from morality, and while they follow their priests in the ceremonies and rites of the church they resent their interference in industrial affairs. The Pole and Ruthenian love freedom and are more self-assertive and independent than the Hungarian and Lithuanian. Among the German Poles are many who read and think; they are tainted with the materialism of the Social Democracy of Germany and their activity and prominence among their fellow-citizens give plausibility to the assertion often made that atheism and socialism prevail among the Sclavs of the coal fields. The Sclav, as well as all other classes of laborers, is not impervious to the fascinating dreams of socialists who promise a plentiful supply of food without hard labor, and their minds, empty of facts to the contrary, cannot resist the

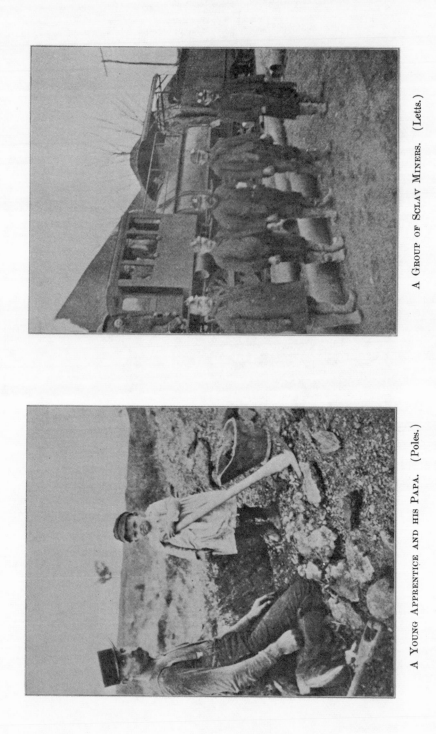

A Young Apprentice and his Papa. (Poles.)

A Group of Sclav Miners. (Letts.)

specious arguments of theorists, so that it is not to be wondered at if they display a propensity to listen to the superficial prattle of those who have sipped at the table of Lasalle, Bebel, Knebknecht, etc. The Sclav, however, is devout and needs the leadership of a Whateley, a Kingsley, a Stöcker or a Ketteler to guide him to a better and nobler type of manhood. He is conservative and slow to deviate from the customs and usages of his ancestors; he is better pleased to follow the old paths than enter upon new ones; and this conservatism of the Sclavs leads us to think that the dreams of socialists will have little influence among these peoples, providing an intelligent and persistent effort were put forth to show them the benefits of the capitalistic system, the limits set to the production of wealth by the forces and laws of nature, the difficulty and hardship which have ever accompanied the work of getting food, and that the laws of the game carried on between capital and labor are, on the whole, just to both parties.

The Sclavs are persistent. Think of the opposition and social ostracism which greeted them when first they came to these coal fields. They were abused in the press and on the platform, maltreated in the works and pelted on the streets, cuffed by jealous workmen and clubbed by greedy constables, exorbitantly fined by justices of the peace and unjustly imprisoned by petty officials, cheated of their wages and denied the rights of civilized men, driven to caves for shelter and housed in rickety shanties not fit to shelter cattle — but the pioneer Sclavs stood it all and by perseverance gained a firmer foothold as the years passed by. This has always been the conduct of men when their economic advantages are invaded. The Anglo-Saxons of these coal fields acted toward the Sclavs invading what they considered their means of livelihood precisely as did their fathers when the Flemish weavers, who excited the bitterest animosity in the minds of the gilds of that day, appeared in England in the thirteenth century. Religious precepts and principles count for naught when the means of subsistence of the average laborer is threatened. But, however great be the opposition to the introduction of new factors into an industry, if the industry needs them in order to render better service to humanity, they will be introduced. The Sclav found standing room in the anthracite industry because he was needed there for the progressive development of the seams of coal, and the Anglo-Saxon employee soon felt that this new hand which he so vigorously cursed, could be used to his advantage. Adam Smith said, if we want the coöperation of some of our fellow-men, "we address ourselves, not to their humanity, but to their self-love, and never talk to them of our own necessities but of their advantages." Mine employees say the operators brought the Sclav into the coal fields to break the back of Anglo-Saxon laborers. That may be true, but it is equally true that the Sclav would never have attained the power he has attained if the employees, because of personal advantages, had not welcomed him.

Inconveniences and hardships are the heritage of pioneers in all lands, and that was the lot of the Sclavs who first came to these coal fields. The generation of Welsh, Irish, English, and Scotch, which is now fast dying away, that came to these inhospitable regions in the forties and fifties of the last century, also endured great hardships. It was not an unusual sight fifteen years ago to see troops of Sclavs leading a communal life in a mining breach, which reminded one of the cave-dwellers of primitive times — only the virgin forest, the crystal stream and the beasts and birds were wanting. Many a conscienceless Saxon made a fortune by sheltering these peoples. Old barns were rigged up with bunks where from forty to fifty men lived and the landlord charged a dollar a month per tenant. They led a happy life, however, for they were a strong people and used to hard living and hard work. When company houses were placed at their disposal, they were the worst in the patch and these were necessarily crowded, for the English-speaking would neither rent houses nor give board to the "foreigners." Many of these rugged Sclavs suffered much from the cold of winter, but the good wages they earned were inducement enough to them to endure hardship and suffering in a strange land. The people, who had for centuries stood the brunt of the conflicts on the continent of Europe when the hordes of the East came as ravenous wolves to devastate the fair plains of the Danube and the Rhine, were not to be daunted by hardships in the mines and inconveniences in dwelling accommodations, if only the supply of daily food was abundant, the flow of lager plentiful and their friend "polinki" near to banish care.

It would be easy to multiply instances of abuses perpetrated on the ignorant Sclav. Sometimes a miner would pay him $1.57 a day instead of $1.75. There are instances where foremen were parties to the fraud, while the method of paying laborers in vogue in the Northern coal fields until lately facilitated fraudulent practices. The Sclavs to-day are more assertive of their rights and few are those who do not keep an accurate account of all moneys due them. When first they came to the coal fields they worked for less wages than Anglo-Saxons; to-day they insist upon the same wages, and the English-speaking employees will cut prices sooner than Sclavs. In the six months' strike under the Susquehanna Coal Company in Nanticoke in the winter of 1899, the men won their case because of the unyielding stand of the Sclavs. It is this quality that makes them good union men. They insist upon regular prices and are eager for good wages. Wherever there is a prospect of good returns for hard work, they are anxious for it. In this they differ from the descendants of Anglo-Saxons. A superintendent of wide experience said: "The native born will not work hard, the Sclav will if you pay him well." After a short experience in the mines they acquire confidence enough to ask for a chamber, and places which have been abandoned by "white men" are readily taken by them. A chamber in one of the collieries in the Southern coal fields, where four men were killed, was shunned by the English-speaking miners, but a Sclav took it willingly. They are gradually becoming efficient employees, while many among them are as skillful as the Anglo-Saxon miners. Their docility and willingness to work hard commend them often to mine managers in preference to English-speaking mine employees. Six Sclavs employed on a job will do more work than four Sclavs and two Anglo-Saxons, for the latter will do little more than supervise the work of the "foreigners," while each member of a homogeneous group works. It frequently happens that the efficiency of the Sclav employee excites the envy of the Anglo-Saxon. A district superintendent of the Philadelphia and Reading almost precipitated a strike in one of the shafts because he gave a contract to two Poles in preference to English-speaking rivals. In the Northern coal fields if the Sclav miner cuts twelve cars of coal to a keg of powder, he is satisfied and says "me all right, boss," but the English-speaking miner expects to cut eighteen cars to the keg, and if he falls below that he asks for allowance. The Sclav is perfectly willing to work from eight to nine hours a day. Many Anglo-Saxons think that "too long." These qualities of the Sclav employee, under the changed condition of the mines, give him the advantage over men of better education and training.

There are withal dark sides to his character. Sclavs are ignorant, clannish, unclean, suspicious of strangers, revengeful and brutal. There are about 50,000 of our people illiterate, most of whom are found among the Sclavs. They are dirty in their homes and the vast majority of them exercise very little thought in their daily task. They have not been taught to think for themselves and hence waste much physical force as well as much coal which more intelligent mining saves. The Sclav seldom leaves his own colony save for his work, and whenever a stranger visits their quarters he is an object of suspicion. In suits at law, if the justice of the peace places the costs on one of the parties to the dispute, there will be no peace until the other also is equally mulcted. In fights they are brutal beyond description. Under the influence of drink they soon return to the unstable nature of their barbarous ancestors. Scenes of cruelty and horror are of frequent occurrence when the secret and suppressed side of their nature is brought out. The Sclav is irritable and easily provoked when in his cups and few feasts close without a fight. Whenever he asks a favor he cringes at the feet of the bestower, but if you ask a favor of him he expects you to do the same.

Many of them are self-assertive and can only be kept within moderation by stern discipline. They ask for favors constantly if they are not harshly refused, while many of them are inveterate thieves. These undesirable traits of character must be dealt with by the stern hand of the law, and local authorities and courts of justice have much to do in the preservation of the peace of these communities by vigorously prosecuting criminals and promptly punishing offenders. The Sclavs come from countries where the authority of the law and the majesty of rulers are ingrained in the people. There is cause to fear that soon after they land in America, these salutary restraints are lost and they imagine that they can by bribery and wire-pulling defeat the ends of justice.* If these men, in whom the brute lies so near the surface, lose respect for law and constituted authority, one cannot but look with grave apprehension to the future of these communities where industrial friction is so liable to occur.

Sclavs Accumulating Riches.

In the town of Shenandoah real estate is at a premium for the reason that on all sides of the borough lines lies territory that is either not adapted to or will not be sold for building purposes. Hence we find in this borough greater congestion than in any other section of the anthracite coal fields. The Sclavs in this town form about 60 per cent. of the population and began to settle there in the early eighties. The high price asked for real estate has impeded the Sclavs in their aspiration for this kind of property, and yet the amount they have acquired is surprising. In 1901, they held 320 properties in the above town which, according to the assessors' valuation, amounted to $300,000. This was but one-fourth of the market value, so that they owned property estimated at $1,-200,000 which was about 25 per cent. of the real estate held

*A young Sclav in Lansford, when he took out his marriage license, asked the justice of the peace if he would kindly give him a paper exempting him and his friends from arrest if, during the marriage feast, they should be involved in a fight. He was willing to pay a good price for that kind of a license, but he wanted it issued under the seal of the court.

by individuals in the borough. In addition to this they have church edifices, parsonages, parochial school buildings and equipments estimated at $120,000, making a total investment in the town of $1,320,000, in fifteen or twenty years.

In the town of Mt. Carmel they form about 50 per cent. of the population. Here in the same year they held 430 properties which, according to the assessors' valuation, amounted to $150,172, or a market value of $700,688. In this borough there are better opportunities to acquire real estate than in Shenandoah, because an unlimited supply of land is on the market. Here also they have spent nearly $90,000 in churches, parsonages, etc., so that over three quarters of a million dollars has been invested in this town. It is estimated that Sclavs here own over 20 per cent. of the homes.

In Nanticoke in the year 1902, 433 properties were held by Sclavs which, according to the value given in the assessors' books amounted to $103,247, which represented a market valuation of $412,988. The tax-collector believed that fully 30 per cent. of the homes in the borough were owned by Sclavs. The value of the homes range from $350 to $7,000, while the average was $953. Here also are churches, parsonages, parochial school buildings, a convent, etc., which were estimated to be worth $110,000. Hence in this town the Sclavs have invested in real estate over half a million dollars.

In Olyphant, where the Sclavs began to settle in the last decade and form now about 33 per cent. of the population, they hold 130 properties which are valued by the assessors at about $40,762 and represent an actual valuation of $163,048. It is estimated that about 14 per cent. of the homes in the borough are owned by Sclavs. Here also they have two church buildings, two parsonages and a parochial building, which represent a valuation of $30,000. So that in the above borough they hold property to the value of $193,048, all of which has been accumulated in the last decade.

Thus, in the four towns mentioned above, the Sclav portion of our population holds over two and a half million dollars worth of real estate which, as near as we are able to estimate, is an average of about $100 per capita of Sclav population in these towns, all of which has been saved in the last ten or fifteen years. In addition to this, vast sums of money have been sent to relatives in fatherland ; thousands have returned, having earned what they deemed a competency for the remainder of their lives, while large sums are on deposit in savings banks. C. S. Weston, General Agent of the Real Estate Department of the Delaware and Hudson said, that no class of immigrants into the coal fields have a better record on the books of the company for prompt payment for the land sold them for building purposes than the Sclavs. Hundreds of building lots have been sold them in the Lackawanna and Wyoming valleys by the company, and less than two per cent. of the purchasers have failed to meet their payments, while among purchasers of other nationalities a much larger percentage of failures occur. Thousands of our Sclavs live in mining camps wholly owned and controlled by the coal companies, and it is fair to assume that if opportunity were given them to acquire real estate they would do so.

This Sclav acquisition of real estate is not local. It is general all through the anthracite coal fields wherever opportunity is given individuals to purchase property, and there is reason to believe that it will continue. The old virtue of thrift is practised by them, and as financiers from among their own number organize building and loan associations and institute banks, they will each year go on adding to their possessions. Our bankers have learned that the Sclavs save money, and almost every bank doing business in the communities where Sclavs form a considerable part of the population engages a clerk from among that people. This thrift of the Sclav and his anxiety to acquire a home should be encouraged, for it is a guarantee of good conduct, industrial efficiency and social order. Private property is one of the best adjuncts to the preservation of law and order and a pledge to the State of good citizenship. Coal companies who unyieldingly grasp every inch of property under their control and refuse to sell an inch of land to their employees, commit a grave mistake. It would be better policy to give these men an opportunity to take root in the soil; they would unquestionably get better employees and the money invested in real estate near the collieries would be a guarantee of steady habits and permanency of residence. In Lansford, the coal company charges high prices for building lots, and the Sclavs, who form about 70 per cent. of the employees, own only 30 properties in the borough.

In this acquisition of real estate by the Sclav there is one detracting feature. Not only do they build new homes, but they also buy the properties of Anglo-Saxons. Many of these latter, hard pressed by Sclavs in business and in the mines, feel that they cannot stand the competition and so sell and get out. Whole wards in the towns above mentioned are fast passing into the hands of the Sclavs, where formerly English-speaking people lived. This results in driving out from our territory the better class of mine employees, pushed out by the aggressive Sclav.

Do the Sclavs Make Good Citizens?

The acquisition of property brings with it also the desire for citizenship, for a property owner desires a voice in the regulation of the town in which he has interest and where he pays his taxes, hence, we find that, coincident with the purchase of real estate by Sclavs, a large number of them become citizens by naturalization.

In the year 1897 the courts of Lackawanna, Luzerne and Schuylkill drafted a new set of rules to regulate the process of naturalizing aliens, making it more difficult and expensive for those who desire to enjoy fully the privileges of Americans. Previous to the change the foreign born could secure his naturalization papers after a residence of five years and the payment of $2 in fees. The residence clause is the same as before but the applicant now must engage the services of an attorney, who moves the court and who publishes the application three times in the *Legal News*. The judge before whom the applicant appears conducts an examination of the alien on the Constitution of the United States, the Constitution of Pennsylvania and the duties of citizens. The nature of the examination depends on the judge, although, it is claimed, that political elections have their influence upon the result. Under this law the applicant for the rights of citizenship pays $3 for advertisement and $5 for the services of an attorney, and if he resides in localities far removed from the county seat he loses a day's work and spends a few dollars for carfare and meals. It is said that it côsts many Sclavs, residing in sections of Schuylkill and Luzerne counties, from $12 to $15 to secure the rights of the franchise. It is very questionable whether this change is for good in the process of naturalization; one thing is certain, it has been the occasion of devising schemes whereby the increased cost may be evaded.

The Sclav in this matter, as in all others which affect his material interests, moves in a practical manner that commends his business tact and condemns his political ethics. The applicants organize into political clubs, and prepare themselves for the examination. When they are ready they wait the time of election until some aspirant for political honors comes around. A bargain is then made; if he secures them their naturalization papers the club will vote for him. In this way a large number are pushed through, previous to the elections, at little expense to themselves. The courts' regulations are salutary in their effect as far as they compel these aliens to qualify themselves in the form and principles of our government; it is injurious in forcing them into organizations which watch the political horizon and to begin their political life by selling their votes. The first lesson taught these men in the exercise of the franchise is that it is property having market value, which they sell to the highest bidder. To construct a pecuniary barrier across the alien's path to naturalization simply appeals to his cunning, while it gives members of the legal fraternity a lever whereby they may secure votes for themselves or their friends.

Politics has always played an important part in the work of naturalizing aliens. Back in the seventies of the last century, it is alleged that the seal of one of our courts was taken out and aliens, without the aid of judge or lawyer, were made citizens by the score. This highway method is not practised to-day and yet methods are still in vogue not in harmony with the patriotic aim of elevating the standard of efficiency and intelligence of aliens. It is a fatal mistake to teach Sclavs on the threshold of citizenship the crime of venality and that their vote has a commercial value.

Some of the Sclav priests are very active in training their people in the necessary qualifications for citizenship. Those of the Little Greeks have prepared a pamphlet containing the Constitution of the United States in both English and Little Russian in parallel columns with notes explanatory of the text. The efforts of the spiritual leaders are not in vain, for many Sclavs pass a very creditable examination which, in some instances, has received the commendation of the judge. An examination of the naturalization dockets shows that the percentage of failures among Anglo-Saxons is not much lower than that of the Sclav. This may be due, however, to the leniency of the judge, one of whom said he was heartily ashamed of his work after every sitting in a naturalization court. When from 200 to 300 aliens are naturalized in one session, the examination of each applicant cannot be very rigid.

The following table gives the number and percentage of Sclavs and Anglo-Saxons naturalized for three years in the Court of Schuylkill county:

Year.	Sclavs.		Anglo-Saxons.	
	Number.	Percentage.	Number.	Percentage.
1899	257	77.88	73	22.12
1900	459	91.99	40	8.00
1901	324	93.10	24	6.90

Thus out of a total of 1,177 naturalization papers issued in the above county in the years specified, 88.36 per cent. were issued to Sclavs.

The table on page 45 gives the nationality of the persons made citizens.

In Luzerne county, in the year 1900, of 253 naturalization papers issued by the County Court, 173 or 68.37 per cent. were taken out by the Sclavs. In Lackawanna county, in the

Nationality.	1899.	1900.	1901.
Austrian	23	71	119
Italian	7	5	42
Pole	119	2	100
Russian	9	361	11
Hungarian	4	5	2
Lithuanian	95	12	50
Syrian	0	3	0
Chinese	1	0	0
Irish	16	6	7
German	36	12	5
Welsh	3	9	1
Scotch	1	2	1
Swiss	2	1	1
English	12	10	7
Swede	2	0	1
French	0	0	1

same year, the Court of Quarter Sessions issued 371 naturalization papers and the County Court 328. Of the total 699, 401 or 57.37 per cent. were issued to Sclavs. The percentage of Sclavs naturalized in these counties is less than in Schuylkill, but there is reason to believe that many German Poles in the last-mentioned counties are classified as Germans. If we add those classified as Germans in Lackawanna and Luzerne counties to the naturalized Sclav, then, of a total of 952 papers issued, 753 or 80 per cent. would be to Sclavs, which is near the percentage in Schuylkill county. It is frequently the case that German Poles pass as Germans in the Northern coal fields, being anxious possibly to be esteemed as standing on a higher social elevation than is generally awarded the Sclavs.

This aspiration of the Sclav for the full rights of citizenship will continue, for it is purposive. There are many brilliant young men rising among them who cherish political ambition, and they successfully lead their fellow countrymen to acquire the rights of citizenship in order to enhance their prospects and powers in both municipal and county politics. They are gradually appropriating more and more of the spoils of office in municipalities and their power in county elections is annually increasing.

These people have both physical and intellectual qualities which will enrich the blood and brain of the nation, but the political ethics in vogue in our State are far from possessing a character likely to strengthen and elevate the moral nature of the Sclav. His leaders teach him cunning and give him samples of fraud and sharp practice which he is quick to copy. Venality is the common sin of our electors and the Sclav has been corrupted in the very inception of his political life in his adopted country.

Clinging to Old Customs.

Representatives of European nations on American soil perpetuate national customs and traditions with a zeal measured only by their national pride. National customs and habits differ widely and the ideas of propriety and decency which prevail among the various nations differ just as widely. Among the confluence of nations in these coal fields the mores of the people inevitably clash and the result is suspicion, misunderstanding and scandal. It requires a degree of intelligence and culture, far beyond that found in mining communities, to understand that national customs and habits are the unconscious inheritance of each generation which, however repugnant and scandalous to peoples of another race, may be practised with impunity by the people themselves. Tacitus thought female virtue low among Germanic women whose paps were visible, but in getting better acquainted with them he was astonished to find them virtuous. The Lapps are dirty, but their respect for women is far in advance of that of peoples more advanced in civilization. The Esquimaux love their wives but they can look upon them carrying heavy burdens without stirring a finger. Each of us is liable to judge peoples of other nations by the mores of the nation to which we belong and hastily pass judgment on that which comes short of our standard. This practice, though natural, leads to gross injustice and misunderstanding.

Social customs among Sclavs differ greatly from those of English-speaking peoples. The female among the Poles will enter a saloon with her male companion and take a social glass with him. Many Slovak women go about the house bare-footed and scantily clad in the presence of men with a naïveté that is their best safeguard of social purity. Many an Italian woman goes to church in company with her husband and a young man who carries her umbrella and shawl. To Anglo-Saxon neighbors there is only one interpretation possible, but ethnologists explain it as a custom prevalent in southern Italy. Sclav women, about to become mothers, never dream of keeping to their homes, and soon after parturition they are around doing their work in the home. It is nothing unusual to see these women up and at their usual task in and around the house, bare-footed, two or three days after the child is born. Sclav girls marry when they are sixteen or seventeen years of age and sterility is rarely found among the wives. There are no spinsters among the people. A priest speaking upon this subject said: "All our people marry, and our women, thank God, have not learnt American ways as yet." Boys are regarded with greater favor than girls in these homes. At the ceremony of baptism a few of the friends are invited and after that they are known as the "Kum." The advent of a male child is generally celebrated by a feast to which the "Kum" is invited, but although these are considered bound to the family by special ties of friendship, none of them is allowed to touch the feast that is spread until the host invites him. The evening is spent in drinking and dancing and sometimes winds up in a fight.

Marriage customs are unique. The bridegroom purchases the wedding garment of the bride as well as the materials for the feast, and provides from $25 to $30 worth of beer and spirits for the occasion. After the marriage ceremony, which generally takes place in the church, the friends of the contracting parties assemble at the house where the feast is spread and where the "polstertanz" takes place. Lippert says that among the Sclavs the "Kum" or "Gevatter" will not allow anyone to dance with the bride unless he first lays a piece of money in her lap. This is the custom here still. Anyone who desires to dance with the bride must lay 25 cents or 50 cents in her apron before he can get the privilege. Pretty brides realize a handsome sum in this manner. A young Polish girl in Shenandoah got $160 * to start life with from those who danced with her. Besides this the bridal pair has another source of revenue by engaging the services of a wit whose duty it is to collect money from the company present which is to aid the young people in starting life. The collector's duty is to address the guests, praise the virtues of the bride, describe the requisites of the home and the prospects before the wedded couple, and then appeal for contributions. The company responds according to its ability and inclination, the contributions varying from $1 to $5. The amount realized largely depends upon the wit of the collector. There lived in Shamokin a genius in this regard whose services were in great demand in Northumberland, Schuylkill and Luzerne counties. The fellow, however, was notoriously immoral and the tolerant Sclavs resolved to cast him overboard. Others were tried as collectors but none approached the ostracized one and the practical Sclavs, when they saw the collections falling off, recalled the scurrilous wit whose jests and drollery never failed to empty the pockets of the bridal guests. It was only by shipping the fellow to his native home that they were able to dethrone him. The married women in the feast take the young bride and put up her hair which before marriage was left hanging. She now passes into the ranks of matrons and is expected to cover her head with a scarf. But here the influence of a new environment is seen; the young Sclav wives discard the scarf and don a hat covered with bright colored flowers and ribbons.

The Sclav is proverbial for his indifference to sickness and death. It is seen in the case of mothers and their children. It is also evidenced in the relation of husband and wife, and between so-called friends. When the husband of a Slovak woman living in Scranton was brought home, having been killed in the mines, the woman denied he was her husband and bolted the door in the face of the bearers. They laid the body

* A Polish wedding generally lasts from three to five days. Contributions also for the privilege of dancing with the bride may exceed 50 cents, which make it possible for a popular young lady to realize a considerable sum. The average sum, however, collected by this custom would not exceed $50.

on the sidewalk and summoned a policeman who was obliged to force the door before the body could be brought into the home. In Mt. Carmel, a Sclav boarder was killed in the mines and brought home. That night a feast was to take place in the house which the neighbors thought would be postponed because of the death of one of the inmates. To their astonishment they heard the sound of revelry and notified the constable. The officer visited the house, and as he was passing through a dark alley leading to the dwelling he stumbled over something. A light was brought and there laid the body of the boarder. They had thrown out the dead and proceeded to enjoy the feast as prearranged. When some of the neighbors remonstrated with them, the reply was "dead Hungarian no good." In the city of Scranton a sick man, boarding with a family of Sclavs, was taken from the house in the dead of the night, placed in the rear of the lot without shelter or any provision for his needs. Some of the neighbors heard the poor fellow groaning during the night and by investigation found him the following morning more dead than alive.

Such callousness is simply barbarous and connotes a low stage of civilization. It cannot be said that these people fear death as much as they dread the expense and inconvenience incident to sickness and death. Upon them, however, a refining and elevating influence is being exerted. A large number of Sclavs to-day are members of religious fraternal organizations and carry insurance of from $300 to $500. This sum paid on the body of the deceased has a quickening effect upon the Sclav's sentiments of respect for the dead, and on the day of burial the organization attends the funeral and carriages are hired after the expensive fashion of Americans. Of course all Sclavs are not equally indifferent to the claims of humanity when sickness and death invade the home. There are some who equal any among the working class in refinement of sentiment and sympathy. They mourn over the dead, summon their friends to chant plaintive dirges in the home, secrete in the casket objects which the deceased dearly loved, and do not forget the few pence necessary to pay Charion when the soul passes over the dark river. In the Greek Church the relatives of the deceased are careful to provide a candle for each one who attends the funeral, for to neglect this is considered a wilful breach of etiquette and a gross insult to the neglected party. Lester F. Ward has excused the Government of Russia because of much of its crudeness by stating that "Russia is simply a few centuries behind the rest of the civilized world." The same may be said of the Sclavs in these coal fields, and if members of nations more advanced in culture will remember that their ancestors were once in the same stage of civilization and that the evolution of compassion and sympathy with human suffering is the growth of centuries, they will look upon the Sclav with greater tolerance and, with a larger hope, work for his elevation.

All the Sclavs drink. The evil of intemperance among the Poles in their native land has attracted the attention of economists and moralists. The claims of temperance appeal not to the Sclav. He looks upon lager exactly as the English do upon tea or the French upon wine. Many drink to excess, but taking them as a whole the percentage of drunkards among them is smaller than among the Celtic races of these regions, while foremen say that they are not absent from their work because of drink half so much as the English-speaking mine employees. Even in his drinking habit the Sclav does not forget his thrift. They frequently club together and buy lager by the barrel. The night is spent in revelry, and from the Sclav quarters comes the boisterous roar of ribald songs which gradually dies away as the debauchees fall into a drunken stupor. Observers say that it is only when lager and whiskey are mixed that the savage brute is aroused in the breast of the Sclav, and on these occasions brawls occur which end in bloodshed and death.

The student of society will consider the social environment of these men before he will censure them because of these irregularities which so frequently darken the Sclav's history in America. Their colonies are largely composed of males far removed from the wholesome influence of home and deprived of the refining influences of womanhood and family ties. They work hard and live on a simple diet. Their lives are spent in the dull routine of daily toil, and upon them is placed the drudgery of the mining industry. Isolated mining camps where Sclavs live as well as their settlements in mining towns, are not penetrated with the varied and brilliant round of enjoyments which come to the Anglo-Saxons. Is it then a wonder that these strong men break the dull monotony which oppresses them by throwing themselves heartily into drinking sprees which sometimes break up in riot, bloodshed and murder? Men, far from home, and removed from the restraining influence of parents and friends, and possessing the means to dissipate, readily pass beyond the boundaries of moderation and indulge in excesses which were undreamed of in the homes of their boyhood. Many of the boys who left the shop, the mines, the store and the office for the camp during the Civil War, to-day suffer the effects of these few years of degeneracy and dissipation, and the social environment of these Sclavs is in many respects similar to that of the boys in blue in the early sixties. Human nature in the Sclav is much the same as in the Anglo-Saxon. President Roosevelt has said of the Westerner: "The backwood people had to front peril and hardship without stint, and they loved for the moment to leap out of the bounds of their narrow lives and taste the coarse pleasures that are always dear to a strong, simple and primitive people." That is what the Sclavs of our towns and villages do, and if we consider their lot we will better understand the rough, turbulent and coarse outbreaks which often shock us who are more favorably situated.

The Sclavs attend church, but they do not observe the Sabbath. They buy, drink, dance, sing ribald songs, play cards, etc., on Sunday without scruple. A good old Baptist deacon who visited the First ward of Mahanoy City, where the Sclavs reside, on Sunday evening, said: "It was terrible; saloons full blast; singing and dancing and drinking everywhere; it was Sodom and Gomorrah revived; the judgment of God, sir, will fall upon us." In this ward are four flourishing Sclav churches and the devout among the Sclavs don't agree with the Baptist deacon. Indeed, if we mistake not the trend of the times in these mining towns, the Sclav's conception of the Sabbath is gaining ground. The saloons of our towns do double the business on Sunday which they do any other day, and it is not the Sclav only that patronizes them. The saloons are full; the churches are not half full — that is Protestant churches, for the Sclavs attend mass and are often seen kneeling on the sidewalk when the services go on.

The Sclav religiously observes the days on which the saints are commemorated and invariably takes a holiday. On sacred seasons of the year, such as Easter and Christmas, they are at great trouble to commemorate the historical events which form the basis of the Christian religion. On Easter, tombs are constructed in churches and a semi-military religious organization associated with the Church assigns quaternions of its members to guard them. Relays succeed each other for a period equal to that during which Christ is said to have remained in the grave. On Easter also members carry baskets laden with provisions to the Church that the priest may bless them and when they are brought home again the families sit down to the consecrated feast. At Christmas time, members of the Church go from door to door carrying emblems of the nativity and recite the story of the miraculous birth. Accompanying them are grotesque figures, representing the enemies of the Church, which add mirth to the visitations. These parties take up collections which are turned over to the priest. On Easter and Christmas a solemn procession is formed, when sacred relics are carried, and the members, chanting, march around the church or along the aisles within the sacred edifice. On Ascension Day branches of trees are cut down and hung over the doors of the houses and around the pictures of sacred personages in the homes. Irish Catholics as well as Protestants laugh at these usages as puerile, but the sneer would possibly be suppressed if they remembered that our fathers once practiced these ceremonies, and that the hand of time alone has stripped us of them as we rise in the scale of civilization. It will do the same to the Sclav and in a shorter period of time than it took us to get rid of them; but the serious question is, what is there to take the place of these ceremonies, which exert a wholesome influence on the Sclav, when they perish by the touch of a higher civilization? Does not the welfare of the rising generation of Sclavs demand a presentation of religion compatible with the higher civilization which they enjoy, which will take the place of the ceremonies which are incompatible with American ways?

As the Sclavs gain in numbers and confidence they give greater publicity to their native customs and peculiarities. Troops of men will, on idle days, amuse themselves by playing a childish game which affords them much amusement. They carry charms and sacred relics with greater publicity than they did in former years. They do not enjoy their frolics and weddings with the same privacy as in the early years of their life in the coal fields, and it is not an anomaly now to find an English-speaking mine employee seeking the hand of a fair Sclav of native birth. Last Fourth of July, a company of Tyrolese paraded the streets of Mahanoy City with a hand-cart drawn by men, in which was placed a barrel of lager. Over it stood a comrade, goblet in hand and crowned with a garland of laurels, singing some jargon, while sitting on the rear end of the vehicle was another fellow with an accordion Along the streets they marched to the strains of music and at intervals they stopped to drink the good beverage they celebrated in song. It was an imitation of the honor paid Bacchus, which was one of the most joyous festivities of ancient Rome.

These quaint customs, imported from across the ocean, are destined to perish with the generation which fondly cherishes them in a strange land. Many of them are signs of a lower type of civilization than that which prevails among us. And to quote Mr. Ward again as he speaks of Russia: "The light will ultimately penetrate that great empire, and in my humble judgment there is no people on the globe more capable of making better use of it." We believe the same is true of these Sclavs. They are a people well worth all it costs of time and money to give them the light of a higher civilization. The light is penetrating the heterogeneous mass and there are signs which promise better things of this people. Under the influence of the public school and the Miners' Union, the national holidays and public parades, public discussion and police surveillance, a free press and free discussion, the Sclav is both consciously and unconsciously rising to the realization of the better and the richer fruits of civilization. Professor Giddings said: "Association, comradeship and coöperation have converted the wild gorilla into the good gorilla and have brought it to pass that, in the quaint words of Bacon, there is in man's nature a secret inclination and motion toward love of others. . . . It is the rubbing together of crude natures that has made fine natures." The Sclav nature is good material to work upon and full of promise. As he comes in contact with Anglo-Saxons and learns their ways, his wants are increased and his tastes refined. As his life expands his reflective powers are made active and he comes "to sift things, to connect events and pass from one thing to the other." The forces which have effected the civilization of man are operative on the Sclav in these coal fields, and they quicken his thought, refine his feeling and give intelligent direction to his will. No one who has studied him in these towns and villages can deny the fact that he increases in material prosperity at a rapid pace. His intellectual activity is also quickened, the greatest apprehension comes from the degenerating influences arising from his social and political environment. These influences can only be met by the coöperation of patriotic and public-spirited men moved to action by the thought presented by Wayne MacVeagh in a recent address: "It certainly would tend to make private property far more secure in America if the less fortunate majority of our population saw us of the more fortunate minority giving courage and time and thought to efforts to solve these problems and others like them, and thereby to lessen some of the evils which in many cases bear so heavily and unjustly upon the poor."

CHAPTER III.

THE THREE CRISES.

1. The Marriages of Mine Employees. 2. The Number of Children Born to Them. 3. The Angel of Death in the Homes.

The Marriages of Mine Employees.

Among the foreign born in the eight counties where anthracite coal is produced we have 33,623 more males than females. Of the Sclav immigrants into our country an average of 70 per cent. are males. It has been shown by the census returns that the majority of male immigrants from southern Europe are workingmen between 15 and 40 years of age. From 1891 to 1900, 74.8 per cent. of the immigrants was classified in the above age group. This accounts for the large number of bachelors of foreign birth in the coal fields.

Among the native born in the eight counties above mentioned, the females outnumber the males by 3,694. This is due to the migration of the male descendants of foreign born parents who, anxious to leave the mines, are forced out of the coal fields, where ambitious young men have very few openings.

Between the Sclav bachelor and the Anglo-Saxon native born or foreign born spinster there is no fellowship. The races are not so far removed from each other but that their union would result in a progeny beneficial to society. The few mixtures which have taken place show an improved stock, and the co-mingling of brain, brawn and blood result in a better type than either of the originals. The chief hindrance to such unions is what Sir Henry Maine has called the "lofty contempt" of a civilized people for less cultured neighbors. This contempt will only pass away with the generation of Sclav immigrants to the coal fields. In the meantime many spinsters, capable of bearing posterity, are living in isolation, for there is no scheme as Mr. Wallace proposes whereby they may be set apart for that work. Many of them seek marriage as Mr. Ward puts it "with all the subtle arts with which men seek pecuniary gain" but the supply of the right kind of men is short and is not enough to go around. In the meantime the Sclav bachelor is sadly in need of a helpmate and would be a better man in every respect if he were to establish a home.

The anthracite mining population, as above shown, is composed chiefly of foreign born and their descendants. The descendants show strong proclivities to intermarry along racial lines. They do not adhere to these, however, with the tenacity of their parents. Descendants of the Anglò-Saxon races intermarry, while many unions of Anglo-Saxons and Germans are contracted. Religious and social ties largely determine the circle of acquaintances of young people and in all of our towns the native born freely associate. The foreign born are generally endogamous as to race. Out of 64 marriages contracted by foreigners in one of our towns in one year in each case the contracting parties belonged to the same race. The same is also true of the Sclav; out of 118 Sclav young men only two of them married Sclav women of a different race from that of their own.

Religion seems to be a greater barrier than race to the intermarriage of descendants of foreign born parents. There are many instances where Catholics and Protestants marry. As a rule, however, such mixed marriages do not turn out well. As soon as children are born, contention and strife begin. The ceremonies of the churches are different and each parent, adhering to the faith of childhood, insists almost invariably that baptism, confirmation or burial be administered according to the rites of its church. Mixed marriages have a demoralizing effect upon the children. They stand between conflicting creeds and many of them grow up to have no creed whatsoever. In a few instances the husband and wife form a compact that the female children follow the faith of the mother and the male that of the father. A woman, living in one of our towns under such a contract, always prayed for girls, and when a male child was born to the home she believed the devil had a hand in it. To intensify the difference existing between parents by a conscious effort to divide the children results only in a divided household, and religion, which ought to be the "bond of perfection," becomes an occasion of strife and bitter rancor.

In the counties of Lackawanna, Luzerne and Schuylkill, where the vast majority of the people are mine employees, we have the following number of persons married per 1,000 population:

Year.	Lackawanna per 1,000 Pop.	Luzerne per 1,000 Pop.	Schuylkill per 1,000 Pop.
1890	16.50	16.08	15.12
1899	18.78	16.86	16.80
1900	15.26	16.34	16.62
1901	18.22	18.78	18.30
Average.	17.18	17.02	16.71

These figures give an average of 16.97 * married persons per 1,000 population in the three counties, or 9.48 marriages per 1,000 population. The normal marriage rate is between 14 and 16 persons, or between 7 and 8 marriages annually per 1,000 persons in the population.

Mr. Spencer says that "each society taken as a whole displays a process of equilibration in the continuous adjustment of its population to its means of subsistence." The law is well illustrated in the above table. The year 1901 was one of the most prosperous years ever known in the anthracite coal industry, and the marriage rate in each county increased. In Lackawanna county the year 1899 was a prosperous one and the marriage rate is high; in 1900, the year of the strike, the production of the mines decreased and so did the marriage rate.

* From January 1 to June 30 this year 1,134 marriage licenses were issued in Lackawanna county, of which 358 or 31 per cent. were taken out by the Sclavs. This gives a marriage rate of 11 persons per 1,000 population in six months. The average age of the Sclavs was: Male, 26.3 years; female, 21 years. That of the Anglo-Saxon was: Male, 25.5 years; female, 23.2 years. This high marriage rate indicates prosperous times in the coal fields.

These figures, however, do not show the marriage rate of a purely mining population. We can get this more accurately by taking mining towns. The following table gives the number of persons married per 1,000 population in the towns of Shenandoah, Mahanoy City and Olyphant, which are wholly dependent on the mining industry.

Year.	Shenandoah.	Mahanoy City.	Olyphant.
1899	18.25	19.13	29.33
1900	21.60	20.44	26.29
1901	28.86	23.52	29.20
Average.	22.90	21.03	28.27

These figures give us an average for the three towns of 24.06 persons married per 1,000 population. The rate for the three counties, provided all the persons married were residents of these counties, was 16.97 per 1,000 population. A certain number of these, however, was from other counties or states, so that the average would be lower if this would be deducted. If we take 16.97 we find that seven more persons per 1,000 population married annually in mining towns than the average in the three mentioned counties. The towns of Shenandoah and Mahanoy City contain 18.9 per cent. of the population of Schuylkill county, but in the years 1899, 1900 and 1901, 42.87 per cent., 49.7 per cent. and 57.7 per cent. of the total number of marriages in the county in these respective years occurred in these two mining towns.

This high marriage rate is due to the Sclav immigrants, most of whom are vigorous young men who leave their homes in the dawn of manhood to improve their circumstances and station in life. Most of these young men make this country their home. Many of them put up simple dwellings in which the young couple begin the conflict of life as husband and wife, and in it as a rule, a vigorous family will soon appear. Goethe's idea of the man worthy of the name — one who "has had a child, built a house, and planted a tree," prevails among the Sclavs. Of the total number of marriages in the year 1886 in Lackawanna county, the percentage of Sclav marriages was

7.88. In 1895 it had increased to 22.82 per cent. and in 1901 it was 31.35 per cent. The following table shows the percentage of Sclav marriages in the three counties of Lackawanna, Luzerne and Schuylkill for the years 1899–1901.

Year.	Lackawanna.	Luzerne.	Schuylkill.
1899	24.14%	28.28%	26.75%
1900	26.03	32.55	29.28
1901	31.35	34.63	33.58

In these counties the Sclavs form about 16 per cent. of the total population, but nearly 30 per cent. of the total number of marriages is among them. If we again take purely mining towns the preponderance of the Sclav marriages as compared with those of the Anglo-Saxon mine workers comes out more clearly. The following table shows this for the towns of Shenandoah, Mahanoy City and Olyphant.

Year.	Shenandoah.	Mahanoy City.	Olyphant.
1899	74.38%	40.55%	56.40%
1900	72.21	47.46	54.00
1901	74.12	48.92	53.60

In Shenandoah the Sclavs form 60 per cent. of the population and have an average of 73.57 per cent. of the marriages; in Mahanoy City they are less than 30 per cent. of the population and have an average of 45.64 per cent. of the marriages, and in Olyphant they form about 33 per cent. of the population and have an average of 54.70 per cent. of the marriages. This tendency to matrimony among the Sclavs must be regarded as a sign of social progress and a pledge to society of better conduct. When groups of vigorous young men cluster together and find their chief relaxation in drink and riot, the peace of society is disturbed and the animal propensity of the men intensified. The sexual appetite is acknowledged to be one of the strongest in man. Aphrodite waits on Bacchus. Men of strong passions inflamed by spirituous liquors fall into bestiality and abominations contrary to nature. These can only be cured in one way. Excesses among these men sometimes lead to depths of depravity and filth known only to physicians. The sexual appetite, when duly regulated, can be the basis of refining sentiments as well as the means of gratifying the desire for posterity. The founding of a true home means the forming of a bond of family relation, the softening and refining of the passions, the guarantee of social progress and the elevation of human character. We may expect these results among the Sclavs who manifest so strong a tendency to found homes, and, in due time, progress towards higher types of domestic institutions will appear among them as among former immigrants into these coal fields.

The following table, showing the classification of the races of the married parties, clearly sets forth the replacing of the Anglo-Saxon and Germans by the Sclavs as stated in the first chapter. We classify the three years' marriages, 1899–1901, in the towns of Shenandoah, Mahanoy City and Olyphant.

	Sclav.	Anglo-Saxon and German.	Native Born.
Shenandoah	1,008	54	339
Mahanoy City	388	76	392
Olyphant	184	101	75
Total	1,580	231	806

A study also of the industrial status of the bridegrooms shows that the English-speaking section of our population is being forced up by the Sclavs. The following classification of percentages in the various occupations is made of 1,436 bridegrooms in Schuylkill county, for the year 1899.

	Sclavs.	Anglo-Saxons and Germans.	Native Born.
Miners	54.50%	40.00%	9.43%
Laborers*	40.47	22.58	39.90
Business and clerkship.	2.64	12.90	11.29
Trades	2.39	21.50	36.78
Professions	0	3.02	2.60
	100%	100%	100%

The following classification is made of the industrial status of 918 bridegrooms who married in the towns of Shenandoah and Mahanoy City in the years 1900 and 1901 :

*The term laborers in this table includes miners laborers and those classes in and around the mines known as company men.

THREE GENERATIONS OF MINE EMPLOYEES.

	Sclavs.	Anglo-Saxons and Germans.	Native Born.
Miners	57.17%	25.48%	7.97%
Laborers*	37.23	34.15	41.10
Business and clerkship.	3.79	18.83	17.39
Trades	1.81	21.54	29.53
Professions	0	0	4.00
	100%	100%	100%

These figures show that the young men of native birth largely enter the better class of occupations and get out of the mines. The ten young men, in the last two towns, engaged in professions, were native born. In the class of laborers, the native born forms a larger percentage than either Sclavs or Anglo-Saxons. We must remember, however, that in this class we have a great variation in the grade of employment in the mines. Some workers, classified as laborers, are skilled men, the risk they incur is small, the labor is not disagreeable and their wages are high. These jobs are monopolized by the native born. Another class of workers, under the same denomination, works hard, incurs a greater risk, is engaged in dirty and disagreeable work, and the wages are small. In these jobs we invariably find the Sclav. In the other percentages in the column we find the contrast striking. While 57.17 per cent. of the Sclav young men were miners, only 7.97 per cent. of the native born were so engaged. But we find that 46.92 per cent. of the native born were engaged in business, clerical work and the trades, while the Sclavs had only 5.6 per cent. of their number so employed. The proportion of the native born entering the trades in the whole of Schuylkill county is an indicator of the aspirations of the descendants of foreign born parents. They rise to a higher grade of employment in the economic hierarchy and thus improve their social status. In conversation with intelligent Sclavs we find the same aspirations in their life, and the industrious and thrifty among them hope for a better economic and social status for their descendants. Competition for the better positions in and

* The term laborers in this table includes miners' laborers and those classes in and around the mines known as company men.

around the mines will become keener as the Sclavs aspire to them. The principle of primi occupantes will favor those now in possession. These will hold them as long as they can and their descendants will have the advantage of heredity, which tends to produce capacity best suited for the work.

The following table shows the average age of the bride and bridegroom in Schuylkill county, for the year 1899:

	Total Males.	Average Age in Years.	Total Females.	Average Age in Years.
Native born..........	924	25.74	959	23.06
Sclavs	371	26.88	353	22.47
Anglo-Saxons, etc...	71	29.05	64	23.97
General average......		26.22		22.95

During this year 79 widowers married, whose average age was 45.38 years, and 68 widows, whose average age was 38.79 years.

If we take again the purely mining towns of Shenandoah and Mahanoy City, we get the following figures as the average age of the bride and bridegroom:

Nationality.	1900.		1901.	
	Male.	Female.	Male.	Female.
Native born..........	26.63	23.86	24.07	21.60
Sclavs	27.33	22.96	27.24	22.74
Anglo-Saxons, etc...	27.64	21.69	30.00	27.50
General average......	27.07	23.26	26.33	22.49

The average ages of the widowers and widows who married during these years in the above towns were 43.39 and 41.43 years respectively.

In the Northern coal fields in the town of Olyphant, the average age of those who married in the years 1897–1900 inclusive, was:

	Male.	Female.
Native born..........................	25.08 yrs.	23.00 yrs.
Sclav..	26.32 "	19.72 "
Anglò-Saxon..............................	26.08 "	22.78 "
General average................	26.69 yrs.	22.49 yrs.

In the three above-mentioned mining towns, the average age of the bridegroom was 26.58 years, while that of the bride was 22.52 years. These figures are very near those of the average for the whole of Schuylkill county.

The average age at time of marrying among the mining population of England is, bridegrooms 24.06 years and brides 22.46 years. The average of females in our computation comes very close to that of England, while the age of the males in our communities is higher by 2.52 years. This is partly due to the scarcity of women among the Sclavs. The men are forced to defer marriage until such time as they can import a wife. It is also partly due to the rise in the standard of living among the descendants of foreign born parents.

The average age of the Sclav bride should possibly be lower. The law of the State is, that no license can be issued to minors save with the consent of the parents. Fifty per cent. of the Sclav brides in Lackawanna county was just 21 years of age—a uniformity which seems to imply evasion of the law. The age of the Sclav woman entering matrimony is largely determined by locality. If she is to be imported, the average age is high; if she is with parents in this country then the rule, as set forth by Krauss for the Sclavs of southern Europe, applies: "Generally the maidens are married after they pass their sixteenth year, wann die Brüste zu schwellen beginnen."

Lippert says that the custom prevails among the Sclavs of southern Europe to conclude marriages at certain seasons of the year. The classification (see next page) of the marriages of Schuylkill county for the years specified shows clearly that the Sclav abstains from wedlock in the months of March and December.

The Sclavs abstain from marriage during Lent and Advent. The Irish and German Catholics do the same. The only variation is when marriage is compulsory. Among the other nationalities religious observances have no apparent effect. In purely mining towns 25 per cent. more marriages take place in the autumn months, when the mines work most regularly, than

Month.	1899.		1900.		1901.	
	Sclav.	American and Anglo-Saxon.	Sclav.	American and Anglo-Saxon.	Sclav.	American and Anglo-Saxon.
January......	47	86	73	62	101	87
February....	29	68	56	70	44	67
March........	2	64	1	60	3	79
April.........	26	107	32	90	56	89
May..........	39	68	46	69	41	67
June..........	33	103	33	138	36	148
July...........	36	53	41	68	44	82
August.......	30	72	30	71	51	62
September...	34	123	44	94	39	95
October......	47	111	33	112	48	100
November...	52	106	19	94	64	100
December...	10	91	13	89	10	93

at any other season of the year. The following table of marriages by the month in Schuylkill shows the same tendency.

In September, October and November we have 869 marriages.
In December, January and February " " 694 "
In March, April and May " " 606 "
In June, July and August " " 708 "

Here the maximum number of marriages in the fall months exceeds the minimum in the spring months by 30.26 per cent. So that not only do the years of greatest productivity in the mines show the greatest number of marriages, but the months of the year when the mines work most regularly show the same tendency. Economic prosperity soon stimulates matrimony and the average mine employee of marriageable age is ready to found a home whenever the prospect of subsistence is bright. Increased and regular wages go hand in hand with increased marriages in these communities, and the rapidity with which the effect follows the cause shows improvidence.

Among those contracting marriage in Schuylkill county in the year 1899, we found seven divorced men and eight divorced women, all of whom were native born. Among the foreign born divorce is very rare, but it becomes more frequent among descendants of foreign born parents, for the reason that the native born female insists upon equality of rights. Geddes and Thompson speak of the self-abnegation of woman in giving up the "morsel of bacon" to the husband and subsisting herself on bread, as the best arrangement for the economic interests of the family. That domestic relation is of common occurrence among husband and wife of foreign birth. But the young wife of native birth will not sacrifice "the morsel of bacon." She insists on her equal share and, sometimes, demands the lion's share of the income. It is one of the results of the doctrine of "the equality of the sexes" and the attempt to place the female on an equal footing with the male in the competitive industrial struggle for daily bread. These teachings bring two results which become yearly more marked: on the one hand, it destroys what Goethe called the eternal womanliness of the female; and on the other, it disturbs the harmonious domestic relations of the family. The simple and healthful homes of foreign born ancestors, in which the Napoleonic principle is practiced "un mari doit avoir un empire absolusur les actions de sa femine," are happier than the homes founded by their descendants where the equal-rights doctrine is enforced.

The marriage rate in purely mining communities we found to be over 20 persons per 1,000 population. This means that annually about 8,000 persons marry, nearly 50 per cent. of whom are Sclavs. Those who occupy a low economic and social status enter it inconsiderately and afflict society with a progeny that is unlikely to rise above the plane occupied by their parents in society. The native born of foreign born parents generally aspire to a higher plane than that occupied by their ancestors in the industrial hierarchy and, consequently, tend to defer marriage until they are able to give the lady of their choice a comfortably furnished home. These also exercise prudence after marriage, and beget few children. The Sclavs found homes and, their status in civilization being lower, they are satisfied with fewer comforts when they begin married life, and practice fewer restraints after. These conditions point to the Sclavs and the shiftless descendants of foreign born parents as the two sources whence the population of the anthracite coal fields will be replenished in future.

The Number of Children Born.

In studying the birth rate of mining communities we are met with the difficulty of getting accurate statistics. In none of the counties chiefly dependent on the mining industry is the statistics of birth accurately gathered. Luzerne county makes the best attempt. Schuylkill county has a lax system that is of little value, while Lackawanna county does not attempt the task. When we remember how important it is to keep an accurate account of the children born to citizens, on which so often depend the rights of inheritance, the laxity in this respect, common to the State of Pennsylvania, is little short of criminal. Boards of health in mining towns have prosecuted some physicians for not making their returns as required by law. The attempt has only resulted in reaping a few dollars from fines while the physicians persist in their indifference. The intelligence and education of physicians assure us that they know the importance to society of an accurate record of births. To ascertain the cause of this neglect we asked one of them why he did not make his returns; he said: "My education cost me dearly, and I won't do that work for nothing." The State pays assessors for the work and expects physicians to do it gratuitously.

In the mining towns also there are many midwives who are not registered, and hence are not known to the officers of the boards of health. These women generally serve at births among the Sclavs, and make no returns. In addition to these reasons, in many of our towns and boroughs there is no local board of health, and no one pretends to gather the statistics of birth.

The counties of Luzerne and Schuylkill pay five cents for each birth returned by the assessors of the boroughs and townships. This, however, is not inducement enough to call forth the necessary diligence to make the returns accurate. All boroughs are not equally culpable in this regard. The commissioners at Wilkesbarre refuse to pay the assessors for the births they return, if they have reason to believe that the returns are not accurate. The following table gives the birth rate as returned in Schuylkill and Luzerne counties for the years specified :

Year.	Schuylkill per 1,000 Pop.	Luzerne per 1,000 Pop.	Year.	Schuylkill per 1,000 Pop.	Luzerne per 1,000 Pop.
1894	25.44		1898	29.86	28.36
1895	26.90		1899	27.40	25.73
1896	26.22	27.20	1900	26.77	27.63
1897	27.42	26.16	1901	28.26	28.40

The inaccuracy of the record is seen in the fact that the returns from Shenandoah, for these years, show only an average of 16.03 per 1,000 population, and those of Pottsville 18.07. Special efforts were made to get accurate returns from Mahanoy City for the last few years. The Sclavs, however, have baffled the attempt, although the returns are more accurate than in the ordinary mining town. The figures are as follows :

Year.	Per 1,000 Population.	Year.	Per 1,000 Population.
1896	29.13	1899	33.48
1897	30.69	1900	35.63
1898	35.84	1901	29.79

In Ashland, where the population is wholly made up of descendants of Anglo-Saxon and German parents, we have an average birth-rate for the last eight years of 24.95 per 1,000 population. The returns from Hazleton, for five consecutive years, show an average birth-rate of 31.92 per 1,000 population.

A visit to the Sclav quarters of any mining town shows how prolific he is on American soil. A physician who had considerable practice among them, said : "Among these women it's a birth every year." A Sclav priest said : "Our women carry their children to the full term and thank God they are not Americanized as yet." All the Sclav women of marriageable age are wives and sterility is rarely found among them. Mining towns have always had their streets filled with boys and girls. One explained it by saying : "It's in the soil I guess." The Sclavs will keep up the record, for their quarters are teeming with children. Hungary and Austria lead European nations in their birth rate which averages 44.0 and 38.6 respectively per 1,000 population. The birth-rate among these peoples on American soil cannot be accurately secured, but we may present some data which proves that they have not lost any of their genetic vigor by emigrating.

The nationalities of southern Europe are loyal to their churches, and the several parishes keep a register of births. From these records we cannot get accurate data to determine the birth rate, for the reasons that the priests do not know the exact number of souls in their parishes. Families and single men constantly migrate to and from the coal fields. The following figures were given us by priests in charge of Sclav congregations:

Number in Parish.	Number of Families.	Number of Births in 1901.	Per 1,000 Population.
1,700	350	140	72.35
1,500	200	110	73.33
8,000	1,300	511	63.87
4,000	500	220	50.00
1,000	251	72	72.00

There is considerable uniformity in the figures as found in parishes far removed from each other and, although they cannot be said to be accurate, they indicate a high birth-rate among these people.

Under normal conditions, natality never falls below 20 to 1,000 population, and never goes over 50. The norm may be fixed at 35, and a vigorous and healthful community has a birth-rate between 30 and 40 per 1,000 population. Natality in Sclav colonies in anthracite mining communities exceeds this, for the reason that most of them are young people, and the age groups below 20 and over 40 years are virtually wanting. Under these relations it is possible for Sclav colonies to raise the birth-rate to 70 per 1,000 of Sclav population. Nitti says that Catholicism, according to the statistics of increase in Europe, is less favorable to increase of population than Protestanism. That is not the case in these communities. All Sclavs practically are adherents of the Roman or Greek Catholic Church, but they multiply as rapidly as any group of immigrants to the anthracite coal fields. Their women regard children as a blessing from God. They are on a plane of civilization where instinct governs propagation, and the economic well being enjoyed in these regions removes all considerations of prudence as to the number of children brought into the world. The high birth-rate of the Sclavs will furnish the anthracite coal fields with a population amply sufficient to man the collieries. These will be physically and intellectually better adapted to the work and will continue, through the quiet process of industrial competition, to replace Anglo-Saxons and Germans in our towns and villages. Professor Giddings says: "Birth rate diminishes as the rate of individual evolution increases." This law will undoubtedly operate on the Sclav, but it is as yet a long way off. Unconscious natality is now the state in which they live, and to pass from that to rational and methodical development will require some time.

The naïveté of enceinte Sclav women is a subject of comment among their neighbors, and the ease with which they pass the crisis of parturition is a surprise to English-speaking women. One of them was seen milking a cow the third day after her child was born, and another was on the culm bank gleaning coal and wading through a creek on the way home with her burden on her back when her child was not a week old. This does not equal what Jukic saw in Bosnia. There a woman of the southern Sclavs was seen barefooted cutting ice in a frozen brook the day after her child was born. The husbands have a word to say on the question of the "Wochenbett." They expect their wives to be at the house work within a week, and if they are not most husbands want to know the reason why.

Anglo-Saxons, who settled in these regions in the first 25 years of the development of the coal fields, were most of them miners in their native country. The fecundity of their women, who are still found in mining towns, gives us the number of children born to women of the mining class. The following table shows the average number of children born to women who were married over 30 years as found in the town of Olyphant. We classify them into three groups, according to the nativity of the parents.

	Average Number of Children Born.	Average Number of Children Living.	Per Cent. of Children Living.
To foreign born parents........	9.20	5.16	56.16%
To parents, one native born....	9.10	5.00	55.00
To native born parents	6.81	4.50	69.45

In the town of Blakely we got results nearly the same. The figures are the following :

	Average Born.	Average Living.	Per Cent. Living.
To foreign born parents.........	9.5	5.1	53.70%
To parents, one native born....	8.7	4.9	56.32
To native born parents	6.0	4.0	73.33

The figures show a lower birth-rate among the native born, but a larger percentage of their children live than of those of foreign born parents. The birth-rate is as 2:3, but the number of children living is as 4:5. So that 13 native born mothers — taking the above ratio — would have as many children living as 11 foreign born mothers.

The early settlers in these coal fields were largely from the British Isles. They came in the dawn of manhood and womanhood and raised families in these communities. It is interesting to learn how large their families have been. The following figures are based on a computation of the women of Irish, English and Welsh descent married for 30 years or more.

	Average Born.	Average Living.	Percentage Living.
English parents..................	10.63	5.93	55.62
Irish parents........................	8.85	6.00	67.74
Welsh parents......................	8.30	4.43	53.45

The birth rate is high in each case, but the Irish children show a greater tenacity of life on American soil than those of either the English or Welsh.

No Sclavs were found among the women married for 30 years or more. Another class of women exists — those married from one to thirty years — and among them Sclav mothers are found. We classify these according to nationality and duration of marriage.

Married from 1 to 5 Years. 108 Women; 59 Per Cent. Sclav.

	Average Born.	Average Living.	Percentage Living.
English	1.50	1.50	100.00
Welsh	1.70	1.20	70.60
Austrian	2.00	1.80	89.38
Hungarian	2.00	1.66	91.60
American	1.63	1.36	83.30

Married from 5 to 10 Years. 121 Women; 70 Per Cent. Sclav.

	Average Born.	Average Living.	Percentage Living.
English	4.22	3.28	77.63
Welsh	4.13	2.50	60.34
Austrian	4.00	3.19	79.76
Hungarian	4.24	3.24	76.42
American	2.90	2.10	72.56

Married from 10 to 15 Years. 81 Women; 47 Per Cent. Sclav.

	Average Born.	Average Living.	Percentage Living.
English	5.00	3.62	72.50
Welsh	5.00	3.70	73.24
Austrian	5.30	3.30	51.74
Hungarian	4.00	3.00	72.46
American	3.72	2.84	76.34

Married from 15 to 20 Years. 73 Women; 12 Per Cent. Sclav.

	Average Born.	Average Living.	Percentage Living.
English	5.73	3.46	60.46
Welsh	5.25	4.10	78.09
Austrian	7.50	4.50	60.00
Hungarian	8.00	5.30	66.60
American	5.13	3.60	70.13

Married from 20 to 25 Years. 53 Women; 3 of them Sclav.

	Average Born.	Average Living.	Percentage Living.
English	9.60	6.10	63.53
Welsh	7.80	4.40	56.46
Austrian	7.60	4.60	60.86
Irish	7.80	6.00	76.19
American	7.00	4.70	66.92

MARRIED FROM 25 TO 30 YEARS. 56 WOMEN; NO SCLAVS.

	Average Born.	Average Living.	Percentage Living.
English	9.00	5.60	62.20
Welsh	7.60	4.76	62.63
Irish	9.00	6.54	72.00
American	9.10	5.50	60.44

The Sclav families are found in the three first groups and the percentage of their children living will favorably compare with that of any other people.

The borough taken by us is wholly made up of foreign born and their descendants. Hence in the last class are found native born women of foreign born parents. These in the three first groups show a tendency to a diminished birth rate as compared with the foreign born women. But in the three last groups no perceptible difference exists. The variation is due to the inroad of ideas of prudence and restraint in procreation among the young women of native birth, while the older generation of native born mothers were dominated by the ideas and sentiments which swayed their parents from across the ocean. How different these sentiments are may be judged from the utterance of one of these younger mothers, who was one of nine children; she had given birth to her second child within three years of her marriage, and under the restraint due to the care of her offspring, she said: "No more for me if I have to go to hell for it." In a mothers' meeting composed of foreign born women and their descendants, the lecturer attempted to show the evil consequences of preventative measures as used by women. The elder mothers thought it sound doctrine, but the younger said: "That's an old song." It is this sentiment which permeates the younger generation of women of foreign born parents that results in the discovery of Dr. G. J. Engelmann, who found 21 per cent. of the women of laboring classes of St. Louis sterile, and those who bore children did not have more than 2.1 to the family. The doctor says that "sterility has gone from worse to worse in the face of gynecologic progress," and comes to the conclusion that the main causes of this are moral and not physical. Guyau has said of France: "En somme la dépopulation francaise est purement et simplement une question de morale." Lilienfeld has come to the same conclusion, that the decrease and stagnation in population are due more to moral than physical causes.

As a conclusion from his studies Dr. Engelmann states: "Greater luxury and wealth go hand in hand with higher sterility." The prime cause of this lower birth-rate among the women of foreign born parents is the craving for social enjoyment. It is the suppression of the maternal instinct by the love of excitement in social relations. Dress, society, amusements — these cut across the sacred instinct of motherhood. Virchow said that the working classes had no pleasure save those they could attain by sexual relations and the excitement of alcoholic drinks. The tension of life is much higher to-day than it was a quarter of a century ago, and the rush of excitement opens various avenues of enjoyment to young people which their ancestors never tasted. They partake of these enjoyments and restrict procreation — a thing their mothers thought a mortal sin. It sometimes proves mortal to the daughters for, with all their skill, nature will avenge herself. Dr. W. H. Wathen said: "It is important to impress upon the laity that an abortion at the end of the first month is just as criminal as an abortion at the end of the fifth month, because if there is even any spirituality in the child it must be at the beginning. Induced abortion is a moral, religious, and physical evil. Induced abortions are followed by diseases much more severe than if they occurred naturally."

The crusade against large families, which Geddes and Thompson describe as "the extreme organic nemesis of intemperance and improvidence," is doing its work among the laboring classes very effectually. But the evil comes in when the illiterate laity practices methods which mean physical and moral ruin. The above authors see this evil and while advocating Neo-Malthusianism they "protest against regarding artificial means of preventing fertilization as adequate solutions of sexual responsibility." They add that "after all, the solution is primarily one of temperance." * Will temperance suffice? Most of the women of the working classes are so fecund that "nolle tangere" is the only safe rule, and, social relations being what they are, that is out of the question. Tolstoi has cried against the bestiality of the sexual relations. Can it be otherwise under existing conditions among the majority of the working classes? Four rooms, scantily furnished, are not favorable conditions to temperance. Ellis il font diaque nuit, said one of the miners. To guard against the natural sequence of this indulgence many crude methods are employed which result in physical and moral ruin. The "fruits of philosophy," given to the illiterate, are the fruits of Sodom, and the advocates of restriction have a grave responsibility resting upon them safely to lead the blind.

The Angel of Death in the Home.

Geddes and Thompson have said that death is the price paid for a body, the penalty its attainment and possession sooner or later involves, and in speaking of the "chain of life" which is continuous they compare the bodies to "torches which burn out, while the living flame has passed throughout the organic series unextinguished. The bodies are the leaves which fall in dying from the continuous growing branch." The eminent biologists very poetically explain the fact of death, but the moment we pass from the realm of imagination to that of the actual life of society and observe how swiftly the torches are extinguished among certain classes of our population, and how soon some leaves fall off certain sections of the "continuously growing branch," the poetry turns to a tragedy replete with suffering and waste. It has long been observed that a high birth-rate among the lower classes is invariably accompanied by a high death-rate. This law is verified in the study of the death-rate of anthracite communities. The mortality among the Sclàv children in towns where sanitary conditions are

* A miner's wife, who had borne 9 children in 12 years and whose health was ruined, protested against the brutality of her husband, when the fellow said: "Well I'll take the pay and go elsewhere."

unfavorable to health and where families crowd into small houses unfit for human habitation, is great.

Some philanthropists have suggested the organization of mothers' meetings to teach the first principles in the care of children to these families. That undoubtedly would do some good, but before a reform can be effected, the greed of landlords must be restrained, local boards of health must be regenerated, and an effectual curb placed on the animalism of human nature which defies the teaching of science.

Gynecologists say that the health of the child as well as that of the mother demands at least two or three years between births. In most homes with large families women get along with half that. They pay the penalty. Nature, taxed beyond the limit of profitable expenditure of force, breaks down. And while this goes on "mothers' meetings" will not check this waste. Loria touches deeper depths when he says: "La duree de la vie d'un homme est essentiellement le resultat de ses conditions de richesse on de pauvrete: et ceta est si vrai que la riche a une existence moyenne de 55 a 56 ans, tandis que la moyenne de la vie du pauvre est de 28 ans."

In the year 1900, the death-rate for the whole of Pennsylvania was 14.3 per 1,000 population, which is evidently too low. The death-rate as computed by the census for a registration area was 17.8 per 1,000 population. We would, however, expect a lower death-rate in the State of Pennsylvania than in a larger area, because of the large number of immigrants in the State, who are in age groups where the expectation of life is high. In the anthracite coal fields many deaths are due to the nature of the industry. In 1901, 513 were killed in the mines or 3.47 per 1,000 employees or about .81 per 1,000 population. Among the 100,000 Sclavs in our area, there are practically no aged people. This ought to reduce our death-rate. We find, however, that the death-rate in anthracite coal communities is higher than that of the registration area. This cannot be accounted for by deaths due to the mining industry, for as was above shown, it is less than 1 per 1,000 population.

The death-rate in the several towns varies greatly. The following table gives that of the places mentioned for the year 1900:

Per 1,000 Population.		Per 1,000 Population.	
Carbondale	21.8	Pottsville	15.5
Mt. Carmel	22.4	Scranton	20.7
Pittston	21.8	Wilkesbarre	16.6
Plymouth	21.0	Mahanoy City	26.7
Hazleton	14.4	Average	20.1

The general average in this table is 2.3 per 1,000 population higher than that of the census returns. In the town of Mahanoy City the death-rate for the last eight years has averaged 21.1 per 1,000 population. In the year 1900, the average death-rate of Luzerne county was nearly that of the State, but in Plymouth, a purely mining town, it was 21 per 1,000 population. The death-rate in the whole of Schuylkill county for the year 1900 was 17.11 per 1,000 population and yet in Shenandoah, in the same year, it was 24.45, and for the last six years it reached an average of 22.23 in that town. The mortality returns from twenty towns in the coal fields in 1901 show an average death-rate of 19.65 per 1,000 population.

A study of the deaths which have occurred in the borough of Mahanoy City from 1894 to 1901 inclusive, gives the following figures for the age groups of the three classes, Sclavs, Anglo-Saxons and native born:

Parents.	Age, 1 Year.		Age, 2 to 5 Years.		Age, 6 to 20 Years.		Age, 20 Years and Over.		Total.		Grand Total.	Percentage of Total.
	M.	F.	M.	F.	M.	F.	M.	F.	M.	F.		
Sclavs	249	189	60	49	22	12	165	34	496	284	780	31.37
Ang. Saxons.	78	61	34	24	46	23	389	292	547	400	947	38.09
Nat. born	215	165	65	53	50	33	96	82	426	333	759	30.54
Total	542	415	159	126	118	68	650	408	1469	1017	2486	100

In this mining town the Sclavs form about 30 per cent. of the population and furnish during these years 31.37 per cent. of the total number of deaths, which is a death-rate of 22 per 1,000 of their population, notwithstanding the fact that the families are young and many bachelors are found among them. This enormous waste of life among the Sclavs is due to infant mortality : 56.1 per cent. of the deaths are of children in their first year of life. Another 13.9 per cent. die between two and five years, making a total of 70 per cent. of the deaths among children before they reach five years of age. If we reduce the above table to percentages, infant mortality as seen in the two first age groups is very high.

Age at Decease.	Sclavs.	English-Speaking.	General.
1 year	56.1%	30.4%	38.4%
2 to 5 years	13.9	10.3	11.4
6 to 20 years	4.3	8.5	7.4
20 years and over	25.7	50.8	42.8
	100	100	100

Among the English-speaking, children not yet five years of age form 40.7 per cent. of the total mortality. Those 20 years and over form 50.8 per cent. The mortality of the aged English-speaking reduces the percentage in the two first age groups in this column. If this were reduced to the same percentage as found in the Sclavs' column, infant mortality among the English-speaking would be 54.1 per cent. In the general column infants not five years of age form nearly 50 per cent. of the total mortality.

The study of the deceased in Shenandoah reveals the same high percentage of infant mortality. During the last six years an annual average of 62.1 per cent. of the total deaths is that of infants not five years of age.

Here then is a tremendous waste of life. No mothers among us work in factories ; the towns are not congested as in large cities ; unsanitary conditions prevail and intoxicants are indulged in, but the chief cause of this infant mortality lies in ignorance. Better sanitary conditions ought to prevail, and if the Christian people of these communities who bewail the slaughter of the infants of Bethlehem were to arouse themselves to the slaughter of the innocents which goes on annually in anthracite mining towns, better conditions would soon prevail and a moiety of this suffering and loss would be mitigated.

During the eight years from 1894–1900, 27 per cent. of the children born in Mahanoy City died before they reached one year of life; if we compute the number dying before they are five years of age, the percentage is 35.04. The still-born are excluded from this computation. The only two countries in Europe which exceed this rate of mortality among infants during the first year of life, are Bavaria and Russia, which have 30.6 per cent. and 29.6 per cent. respectively. The economic condition of the mining population is far better than that of the European countries referred to, and infant mortality, in a locality well favored by nature, should not approach the above figures. The Sclavs are amenable to discipline and local boards of health, were they alive to their duties, would so improve the sanitary conditions of these towns as to remove many evils which now slay the infants. An enlightened philanthropy, which would impart to these mothers information regarding the care of infants and the best food for them in a climate which varies greatly from that of their native homes, would partly check this quenching of the torches. Loria says that among the rich in England infant mortality under five years of age is not over 5.7 per cent. Here it is 35.04 per cent.

The above figures of infant mortality also show how much more susceptible the male children are to the ravages of disease than the female. Of the 1,242 children, which died before they reached five years of age, 701 or 56.44 per cent. were male, and 541 or 43.56 per cent. were female.* Thus of every 100 deaths among children under five years of age, there would be about 13 more boys than girls. Of 99 children which lived on an average of only four days, 60 of them were male, and 39 female. The male child in its prenatal condition seems to show the same feebleness to resist antagonistic forces. A woman of considerable experience said: "It is much easier to kill a male than a female fœtus." Among the still-born the proportion of boys far exceeds that of girls. Dr. Ploss puts it at 100 girls to 140 boys.

* The returns for births being inaccurate, it is impossible to state correctly the ratio between the male and female children born to our mine workers.

In the years 1894–1900, there were 85 still-born births in Mahanoy City, of these 68 or 80 per cent. belonged to English-speaking mothers, and 17 or 20 per cent. to Sclav mothers. The figures clearly show that the native born women of foreign born parents give birth to a larger proportion of still-born children than do the Sclav women. Does this tendency go hand in hand with that of artificial restriction of the birth-rate? It is true that the higher man rises in social conditions the more it costs to perpetuate the race. The question that rises is, does the increased number of still-born children of native born women of foreign born parents show a pathological condition, or is it an incident in the higher evolution of the race? Dr. Engelmann says that relative sterility among those who have conceived but who never have carried a child a full term, is higher among Americans than the foreign born. In the former it is from 9 to 12 per cent. and in the latter from 3 to 6 per cent. This seems to point to pathological conditions as the cause of the increase. The evil is not due to one sex more than another. It is the opinion of those who observe the progress of life in anthracite towns, that the sons of foreign born parents are not so clean as were their fathers.

A classification of the deaths according to the month of the year in which they occurred gives the following percentages:

January	9.7%	July	11.9%
February	7.7	August	10.7
March	8.7	September	8.1
April	7.0	October	7.6
May	6.1	November	7.9
June	6.7	December	7.9

If we divide the year into four quarters we have the following rate:

January, February, March	694 or 26.2 %
April, May, June	526 or 19.8
July, August, September	809 or 30.5
October, November, December	616 or 23.4

The summer months are the most fatal. Then the infant "torches" go out, for at this season the unsanitary condition of these towns shows how fatal filth and uncleanness can be. In the winter months the aged die, and this season stands next to the highest in the rate of mortality. This comes out clearly in the following table which gives the average age of the deceased of each month :

Month.	Female.	Male.	Month.	Female.	Male.
January	27.6 yrs.	28.7 yrs.	July	16.5 yrs.	16.0 yrs.
February...	24.2 "	28.9 "	August	14.8 "	16.5 "
March	27.6 "	28.4 "	September ...	17.2 "	20.9 "
April........	27.5 "	20.2 "	October.......	26.4 "	27.0 "
May	22.1 "	24.1 "	November ...	26.2 "	26.4 "
June.	29.8 "	22.7 "	December ...	26.7 "	30.5 "

The average age of the deceased was 24 years for the female and 24.8 years for the male. According to the last census the average age of the deceased in the country at large was 35.2 years.

From the study of the three crises in anthracite coal communities, one feels that the greatest need of reform is in a sphere where it is most difficult to accomplish. The life and death, the character and usefulness of the individual, which as Sir Henry Maine says is the center of our modern civilization, depend upon the intelligence and character of the parties contracting marriage. As long as the shiftless and the ignorant, the impoverished and the careless, contract marriage and are swayed by animal instinct only in the sexual relation, all reforms must be futile. Society will be as Sisyphus, ever striving and ever called anew to the same task. Romance writers generally associate poverty with a large family. Their observations are correct. Everywhere in these mining towns the shiftless and intemperate bring many children into the world, whose bodies are not properly fed, whose minds are distorted and whose hearts are cursed. The parents have nothing and, falling into a state of indifference as to the morrow and its evils, they inflict society with their progeny, and scotch the advancement of man. The law of social capillarity has no ameliorating influence on them. Their envy is excited as they see others rising in the social scale, who once occupied the same social status as themselves. As immigrants they started life in a new environment on a par; by thrift and industry, some accumulate wealth, others accumulate nothing save misery and vice, which are intensified in a manifold progeny. The deterioration of the shiftless is rapid. From among them rise the disturbers of industrial peace. They will not work themselves and will not permit others to do so in peace. They have a natural repugnance to manual labor, and go around to persuade others who do so that they should have ease and plenty. It is the source of class distinction, and the root of it lies in personal qualifications and capacity to carry on the conflict of life.

Under the laws of nature the brood of the unfit would soon be eliminated, but under our social relations they are fed, clothed and educated free of charge to the parents, who consume all they make in gratification of appetite. What to do with this class baffles the wisest reformers. It may mean hardship to apply to it the doctrine of Nietzsche, and yet some means should be used to eliminate this parasitic class from the social body.

On the other hand, we have the vast majority of workers who are operated upon by the law of social capillarity. They rise in the social hierarchy. Their wants multiply, and as they advance in luxury and wealth, their sterility increases. Professor Giddings says: "When the entire population voluntarily diminishes its birth-rate, it gives indubitable proof that it severely feels the pressure of its natural tendency to increase faster than it is possible to raise the general plane of living." The descendants of the foreign born raise the general plane of their living and it is at the cost of lowered natality. The productivity of the mines is at its meridian. Each year these collieries draw nearer to the point of marginal productivity, and in the natural decline of the industry the share given labor of the productive wealth will not be increased. Hence if members of the social group raise the standard of living it must be at the cost of natality. The great law that individuation and genesis are antagonistic is well illustrated here. The income of each family in our communities is limited and if it is spent in raising the standard of living, then natality must be lowered.

But then the question comes, is this raising the standard of living and lowering natality, synonymous with social progress? Lilienfeld says that it is the law of general validity in biology, that the higher living beings raise themselves in the hierarchical scale of organism, their prolific energy grows weaker. The same truth has been emphasized by Spencer who says, that among the lowly-organized creatures, mortality is enormous, and longevity and diminution of fertility are found in ascending to creatures of higher and higher developments. In biology this is true, but unfortunately the laws of biology are not followed by men endowed with intelligence to choose life or death. In communities enjoying a high degree of culture, a low natality may be due to morbid conditions superinduced by prostitution, disease, degeneracy and derogation of the sexes from their respective types. When the medical fraternity discussed the paper of Dr. Engelmann on "Sterility among American Women," one of the speakers said that gonorrhœa was responsible for the pathological conditions genecologists were called upon to treat. When a low natality is due to disease, destructive forces are at work, and to attribute it to a high degree of physical, intellectual and moral perfection of the people is to cherish false hope and flatter the vanity of degenerates blind to their own retrogression.

A large army of men and women, anxious to do the right and conscientious in marital relations, ask a solution of the question, is it right to raise the standard of living at the expense of natality? On one hand stands the "nemesis of intemperance and improvidence" which threatens to devour all the substance of the parents; and on the other hand great sterility and a diminished natality which threaten depopulation and degeneracy. Mr. Ward says: "Just as every one is his own judge of how much he shall eat and drink, of what commodities he wants to render life enjoyable, so every one should be his own judge of how large a family he desires, and should have power in the same degree to leave off when the requisite number is reached." But the trouble is that most men have not the power to leave off. How many of the poorer classes can say as once was said, "C'est le seul plaisir que nous puissions nous passer." A miner above the average in intelligence said: "Abstinence is out of the question; among most men the sexual appetite is stronger than the craving for food." The sentiments which prevailed among men who purchased their wives still exist among many in the lower stratum of society. The woman is something for their use for which they paid the price. When women, raised in a social environment which shatters the traditional views of the old world regarding marital relations, revolt against the "wife-by-purchase" code of ethics and resort to artificial means to prevent conception, they go to the drug store or call on the doctor and end as physical wrecks. Population, we all agree, should be restricted. Restriction is most needed among the classes where increase is now most rapid, viz., the shiftless and thriftless. And among them the disciples of Neo-Malthusianism will preach temperance and artificial restriction in vain. Artificial restrictions as practiced among the working classes work disastrously. The evils fall upon the women. The physical wrongs they suffer are many, but the moral poison instilled into the genetic source of the race is still more fatal. The question of natality concerns the women more than the men, and the psychical side of this whole question is more important than the physical.

The progress of society is from the physical to the psychical. The pen is mightier than the sword in modern civilization, and the only hope of society is in increased psychical power in the individual. Nature has evolved a high type of organism by the law of selection which is as ruthless as it is mighty. Men have applied artificial selection to plants and animals and the results are gratifying. Mr. Ward is of the opinion that man can only claim true distinction from other animals when he has the courage to apply it to himself. With increased individuation as the goal of civilization and the doctrine of laissez faire so vigorously insisted upon in this relation by the working classes, how can artificial selection be ever exercised in human society? Mr. Wallace's selection of females capable of bearing children, and Dr. Francis Galton's certificates of competency to young people to marry, may result in an improved type of men and women, but what of the teeming throng which, while experiments are being made, multiplies so rapidly that the selected type will ever be in danger of being swamped by the mass? Nature has her own way to effect these things and when society has the courage to follow more closely her ways in natality, many of the evils which now trouble society will be eliminated.

Specimen of a House built by English-speaking Miners.
(The porch only was built by a carpenter.)

Specimen of a House built by a Sclav Miner. (Ruthenian.)

CHAPTER IV.

DIFFERENT WAYS OF LIVING.

1. How Much Does It Cost to Furnish a House? 2. What do Mine Workers Spend on Clothes? 3. The Money Spent on the Table and on Amusements. 4. The Effect of a Rising Standard of Living.

How Much Does it Cost to Furnish a House?

Adam Smith said that "a man of large revenue, whatever may be his profession, thinks he ought to live like other men of large revenue and to spend a great part of his time in festivity, in vanity, and in dissipation." This well illustrates the law of imitation upon which Tarde has based his social philosophy. There are many instances of it in mining towns. Mine employees to-day build better homes and furnish them more elaborately because of the example set them by their neighbors. This desire for more elaborate houses and more costly furniture does not always consult integrity and honesty. A street in one of our towns, largely composed of low, unsightly and uncommodious houses, is now graced by modern structures. Before the majority of them was put up, however, fire consumed the uncomely homes and insurance companies protected themselves by cancelling fire insurance policies in that neighborhood. Better furniture is bought and a larger book account is kept by the furniture dealer.

A large number of good dwellings has been erected in mining towns in recent years, which has placed at the command of mine employees a greater variety of abodes. The standard of living of the various nationalities is reflected in the kind of house they rent, the way they furnish it and the means whereby it is heated. As a rule, the Sclavs rent the cheapest houses, furnish them in the scantiest manner and pay very little for the fuel consumed. The English-speaking pay the higher rents, liberally patronize the furniture, hardware and crockery stores, and pay for the coal used. There are exceptions. In one of the mining camps of Schuylkill county, we found an English-speaking family tenanting one of the poorest structures in the patch. There were five in the family that occupied this frame building of two rooms, over which was a low attic lighted by a 12″ x 16″ window. The roof was rotten and the rain and snow came in freely. There was no plastering on the sides of the rooms, and the wall paper, which clung to the walls in defiance of successive showers, bore upon it the marks of many waters. The wife said: "It's a horrible place." The rent was $2 a month, and the lord of this miserable abode brought his wife and little ones into it because the rent was "cheap" and his appetite for strong drink costly. On the other hand, some of the most commodious houses erected by mine employees in the town of Nanticoke were put up by the Poles. The Sclavs do not in all instances voluntarily take the cheaper houses. They are considered socially inferior to the Anglo-Saxons and Germans, and where there is a variety of dwellings, the first choice is given the English-speaking, and what remains goes to the Sclavs. In a mining camp in the Southern coal fields, a dilapidated and ill-built section was assigned to the "foreigners," while the better houses were occupied by Anglo-Saxons. The Sclavs complained bitterly of their houses, and they had just reason to do so. They said, "Me pay more for good house," and most of them would gladly move to better houses were they available. It is the same in our towns. The most dilapidated sections are assigned the Sclav. He is driven there by social and racial discrimination, and it is only when houses are vacant and no "white tenants" can be secured, that the most respectable and most ambitious Sclavs are able to move into better homes and be surrounded by a purer and healthier environment. In towns where opportunity is given them to buy real estate, many of them secure homes of their own, which are healthier and better than the dwellings found in the leprous sections of mining towns.

In every section of the coal fields company houses are found. These vary greatly. The ones recently built by some of the companies are fit habitations for workingmen. Those which were built in the early years of mining, however, are not so. Many of them are miserable shanties unfit for human beings to live in, repulsive to all sense of decency and unfavorable to the cultivation of domestic virtues. Of the number and character of company houses throughout the coal fields we shall speak in a subsequent chapter.

The willingness of men to undergo sacrifice in order to secure a home was never better illustrated than in these mining towns. This tendency of thrifty mine employees manifests itself in all races. In the first half century of mining, a large number of Anglo-Saxons and Germans secured homes which they still tenant. The thrifty Sclav shows the same tendency and, in all towns where opportunity is given the workers to acquire real estate, these people in large numbers either buy or build homes. In Lackawanna county, a number of English-speaking miners working only half time resolved to build their own homes. In two summers they put up six houses. The material in each cost about $350. They were substantial five-roomed houses, which gave ample accommodations and good shelter to the families. They bore traces of crude workmanship, for all the masonry, carpentering and plastering were done by them. The houses, however, answered the purpose, and if John Ruskin were a witness to their efforts we feel confident the company would be admitted into his St. George society. We have seen Sclavs do the same thing, but with greater dispatch and a smaller outlay. A young married Sclav couple resolved to begin life in a house of their own. The bridegroom on Saturday morning ordered lumber, etc., for a house. All the material was delivered that afternoon and laid on the lot he had purchased from the coal company. By Sunday night a house warming was given the young couple in their new home and a keg of "bock" was drunk on the occasion. All told, the two-roomed house represented an outlay of about $125. It was the home of a sturdy Sclav couple resolved upon making the best of their opportunity. The men who believe that the "man who is content with corn, bacon, and a one-roomed cabin has no place in the modern industrial system," would better keep an eye on these people, for they are earnest and will count in the quiet conflict waged in industrial centers. During the last decade these men, who live in crude structures, are the ones who have the largest deposits in the banks. Is not the boastful and pernicious doctrine that "a cheap coat makes a cheap man," working mischief? There was another doctrine in vogue among the men who laid down the foundation of governments. It was "the coat does not make the man." What a picture Carlyle gives us of George Fox when the world was making a "god of its belly." This simple shoemaker, making himself a suit of leather, is ridiculed by the élite who lived in grand houses and considered "the Belly and its adjuncts the grand Reality," but he, the man with a spirit in him, leaves his mark upon the world. Our State of Pennsylvania, before it fell into the hands of men who believe "a cheap coat makes a cheap man," was a monument to the social worth of the man clothed in that suit of leather. Men ridicule the houses in which Sclavs live, but these men are resolute and frugal, and, if signs deceive not, the man in the two-roomed cabin will come out first in the economic contest now being waged in these coal fields.

On the Hazleton mountains one of the companies rented houses for $7 a month. English-speaking people paid that. The Sclav would not. Two families of these immigrants take the house, divide it as best they may, and each pays $3.50 a month. In the same region we found a three-roomed house—each room measured about 12 by 13 feet, where a husband and wife, two children, and 10 boarders lived; 14 souls in all: 4.6 persons to a room. The party lived a communal life. The woman did the purchasing for the company and, at the close of each month, each paid his pro rata share of the total bill. In another house of four rooms, not one of which was more than 13 feet square, there were 14 boarders. When the "boarding boss" was asked how they all managed to get sleeping room, he replied that some of the men were working night, so that these could occupy the beds vacated by the "day men." In a mining patch wholly occupied by Sclavs, where there was great crowding, we asked the men how they managed in summer? They said they slept out in the open air. Of course, there were foes without — mosquitoes — but these were not so troublesome or numerous as the foes within. A Sclav who slept out in summer time, made his bed on the roof of the house. His repose was not sweet: he rolled off his couch and fell on the edge of a picket fence and fractured three of his ribs.

Instances of Sclavs crowding into small houses are not so numerous at present as ten years ago. Of 153 families in a mining town, 111 or 72.61 per cent. had no boarders. The remaining 42 families had from 1 to 7 boarders. More houses are now placed at their disposal. But, to-day, many Sclav families are anxious to get a few boarders for it reduces the item of rent.

The Anglo-Saxons as a rule stand in striking contrast with this. They want no boarders and wish to preserve the hearth sacred for family use. If they pay rent it is from $5 to $9 a month; which is a hundred per cent. higher than the sum paid by Sclavs. The thrifty English-speaking miners, who have lived here for the last quarter of a century, have houses of their own, which are, from a workingman's standpoint, commodious. Houses rented to Anglo-Saxons are subject to the universal law of supply and demand. Mine employees cannot pay more than from $7 to $9 a month rent, and the best kind of a house an investor can build for them must not exceed an outlay of from $1,000 to $1,300. Houses built on this scale contain from 6 to 8 rooms and in them the descendants of the earlier settlers from Great Britain and Germany live. These native born mine employees, starting domestic life by paying $7 a month rent, find that an income of between $400 and $500 a year makes it impossible for them to save the means to purchase a home of their own. Hence in recent years, a far greater number of Sclavs than Anglo-Saxons builds houses and, because of this, the "foreigners" take a firmer hold of the soil than the native born.

As a rule, it may be said that in mining patches as well as in mining towns, the Sclavs pay from $2.50 to $5 a month rent, while the Anglo-Saxons pay from $4.50 to $9. Hence, in the matter of house rent, the cost of living to the English-speaking is a hundred per cent. higher than to the "foreigners."

When we come to the item of house furnishing the contrast is still greater. The native born youth, when he starts life with the lady of his choice, rents a house of six rooms and spends on an average $25 a room in furnishing it. Sometimes, a young couple will be satisfied with four rooms for $5 a month, but never will they begin life below that level. This class then spends in furnishing a home from $100 to $150; very rarely will the bill reach the $200 mark.

The young Sclav, if he rents four rooms to begin life in, expects to take boarders. Many of them begin in two rooms. The average amount which he spends on each room is about $12 and, in many instances, the whole bill does not exceed $35; if it goes up to $50 it reaches the maximum.

A comparison of both homes reveals the difference. In the houses of "white people" the front room is carpeted and comfortably furnished. Here they entertain their friends. In the next room, which is generally large and serving as a kitchen and dining room, the floor is covered with rag-carpet and a large strip of oil-cloth or linoleum under the stove. The cooking stove and all cooking utensils are new — nothing else will do for "young America." A plentiful supply of crockery, a dining room table and half a dozen chairs, give the room a comfortable appearance. The stairs leading to the second story are generally carpeted. The front bedroom is carpeted and furnished with a bedroom suite of "eight pieces." One other bedroom will generally contain a bed, so that the family may entertain a friend in case of need. The third bedroom — a small room generally — is used for storage. Add a heating stove, and a home, where the average native born young people of mining communities begin life, is complete.

The Sclav discards carpet and oilcloth. None of it is seen in the majority of houses. If a few strips of rag carpet are used, it is a sign of an advance above the ordinary racial standard of living. The cooking stove is generally bought at a junk shop. The cooking utensils are few and tinware often serves as a substitute for crockery. A common kitchen table and chairs to match complete the furnishings on the first floor, if made up of one room. If there are two rooms, then the front room has one or two beds in it; no carpet and no bedroom suite of "eight pieces." When shown one of these rooms, we had to sit on the trunk of one of the boarders, for there were no chairs there. The room or rooms on the second floor have beds in them and a few trunks. If a heating stove is purchased, it is the old-fashioned bell-shaped kind, bought second-hand, which is a good heater, and the practical Sclav wants heat and not nickel-plate and polish. All here are articles of necessity, not a trace of luxury seen anywhere. Of course there are exceptions. German Poles, who have been in the country for twenty or thirty years, live differently. We describe the home of the average Sclav which may be found in every town in the anthracite coal fields.

This difference has an important effect upon the economic life of these two classes in our towns and villages. "Young America," with his comfortable home, almost invariably starts life with a debt; the Sclav, with bare necessities, starts free of debt. These are a few examples of the former class from the books of a furniture dealer: Bill, $104.85, paid $20; bill $105, paid $18; bill $100, paid $20; bill $160, paid $45. The balance they agree to pay in monthly installments. The following was an exception: Bill $178.25, paid $150.

The Sclav does business very differently. His bill seldom reaches $50, but he pays it. The rule among them is never to start married life with a debt. The difference soon tells. The English-speaking is handicapped by that debt. Before it is paid a child comes into the home. This extra expense, eating into meagre earnings, still further intensifies the struggle, and the result is that the family lives from hand to mouth and finds it necessary, in crude ways, to practice the doctrine of Neo-Malthusianism. The Sclav, on the other hand, free of debt, soon buys a lot, puts up a simple house, and under that roof, where plenty of coarse food is found, Neo-Malthusianism has no place. Children are born in rapid succession and, though their manner of living may be repugnant to our tastes, yet it must be acknowledged that they live close to nature.

Which of these two classes begins life the better? Which is the better way, to raise the standard of living and mortgage the future years of the wage earner, or apply the principles of common prudence and honesty and regulate one's life according to his prospective income? The craving for elegant houses and elaborate furnishings disturbs the peace of these wage earners, troubles the domestic felicity and destroys the purity of many homes. When Thomas à Becket strewed the floor of his hall with clean hay or rushes in order that the knights and squires who could not get seats might sit on the floor to eat their dinner, were they less valiant fighters in defense of their lord because of that? When the old tavern keeper at Dunfermline slept in the marriage bed of James I., King of England, did he sell better ale and serve better dinners because of that? The social worth of man is the question. Here we have an industry which requires hard muscle and strong brawn, and the man who sleeps free of debt on a simple bed and eats coarse food out of a tin plate, may have greater social worth in the mining industry than the man who sleeps in a bedroom with "eight pieces," eats his food off imported chinaware, and has a debt and a heartache.

Turn now to the fire in the house. In savage times it was the woman's task to care for the fire, and she did it. The Sclav women do it. We have seen native born girls in these mining towns assuming the charge of a home, who could not keep fire in the stove. The "foreign" women get the fuel and keep the fire. They go barefooted to the culm heaps and glean, and, staggering under their burdens, they keep the coal bin full so that nothing is taken from the husband's wages for fuel. Around the breakers, on railroads, or wherever new buildings are erected, the children of the "foreigners" are seen, gathering fuel and tugging it home with a tenacity that is admirable. Some of these women wield the ax with a dexterity and strength that would put fifty per cent. of our men to shame. The English-speaking buy their supply of coal. It costs them from $30 to $35 a year. Many Sclavs do not spend $5 a year on coal. In the due-bills of Anglo-Saxons the item of coal comes in with great regularity; in the Sclavs' due-bills it is rarely met with. Some of the coal companies prohibit these people from gleaning on the culm banks, but it is hard to stop them. In summer time many of these women are up before dawn, and, ere the foreman appears, their load of fuel is safely stowed away in the coal-bin. Sometimes they are "pulled" by the police or detectives. Nevertheless the practice continues. Some of the coal companies have considerable trouble in keeping their coal cars intact on their way to market. The policemen of the Reading Coal and Iron Company are kept busy at Kohonor Junction watching the coal cars, and one of them said: "These Poles are born thieves." There are others besides "foreigners" guilty of stealing prepared coal. A foreman said of an English-speaking family, living in close proximity to the shaft in his charge, that during the last ten years they had not paid a cent for coal. They did not glean it from the bank. They took it from the pockets in the breaker. One of the workers in that home was very zealous about the weighing of the coal, and explained how the poor miner is robbed. Many men in the Middle and Southern coal fields are guilty of "robbing the breaches," or taking coal from the out-crops. These, when caught, are punished.

Thus in the items of house, furniture and fire the Sclav on an average spends less than half of what the Anglo-Saxon and Germans spend.

What do Mine Workers Spend on Clothes?

Carlyle said that the dandy asks for one thing only — "the glance of your eyes." These dandies have always been in the world, and their vanity has been an important factor in civilization. The first steps in clothing were taken in dandyism, and the motive power was egoism before man knew that he had an ego. To-day, woman is more given to the art of exhibiting clothes and adornments than man, but Lippert assures us that it was not so in the early history of our race. He says: "Unter den emfachsten Verhaltnissen der Naturvolker fallt nur ein geringer Anteil am Schmuck auf die Frau; es ist der Mann, der sich am reichsten und auffalligsten schmuckt." The male, leading a more active and energetic life, sought distinction. Decorations offered him the most feasible way to gratify his vanity and herald his deeds of heroism. Hence, when the savage went to war or sat in council, he carefully selected his adornments. Indeed, how many men are there at the present stage of the world's progress wholly exempt of this weakness? A modern philosopher said that all men are equal if stripped of their garments, and the consciousness of that makes most of us hang on to our paraphernalia. The aim of civilization is increased individuality which is attained by the development of the psychical elements in man. All savages are pretty much alike, but a Goethe, or a Darwin or an Agassiz is marked by an individuality which makes him unique among the millions of his contemporaries. The men who possess greatest individuality are furthest removed from the savage whose chief aim was external adornment. Civilization has not gone very far yet, for much of this primitive custom is still in the world. It is seen in the arrangement of the hair, tattooing, conspicuous and cheap jewelry, so common among the working classes, as well as in the colors and feathers displayed by lodges when parading the streets of a mining town.

Civilization, however, is gradually changing this. Woman now pays more attention to dress than man, and adorns herself with glittering jewels or strass. This was well illustrated by a Sclav who said: "Me pay for frau hat $5, me hat 50 cents." The man said it with a gusto that resembled the air of the American who says that his wife is the "best dressed woman in town."

All nations vary in their costumes and among all of them dandyism is found. The Fuegian who chooses beads and glittering trinkets when naked in an arctic climate, is no different from the men and women in these coal fields, who go short of the necessaries of life in order that they may spend more money on trinkets and baubles. Parsimony is never exercised by them. All they care for is to live in a vain show by following the fashion and wasting their substance in cheap dresses and paste jewelry. They live from hand to mouth, and, in hard times, fall into vice or ask for relief. Their wages are spent for articles of no social or individual utility, and they at length fall into hopeless servitude and poverty. The practical Sclav will not, as the Anglo-Saxon, waste his earnings in useless vanities.

The change in the standard of living is reflected in clothing as in every other sphere of human life. Lippert tells us that the inferior copied the superior in ancient society and the same practice is seen in the life of to-day. The improvements in the arts are such that the wife and daughter of the wage earner can purchase a dress at very moderate cost now-a-days which, a century ago, could only be secured by the rich. Consequently, it is nothing unusual to find the daughters of mine employees wearing articles of apparel which, even under our improved system of production, are expensive and absorb a larger portion of the wages than the workmen can well afford. The girls are able to get them because of the parents' sacrifice and indulgence. Imitation of the rich is not only confined to the material of which their dresses are made but also in the multitude and variety of their dresses. The daughters of workingmen have a wardrobe to exhibit, and the craze for variety in dress among female descendants of Anglo-Saxons in mining towns is so great that the Sabbath day is more a dress-day than a holy-day. Does this rage after fashion promote the moral well being of women? Have these daughters, with full wardrobes, greater social worth than their mothers? Lippert says that when the more elaborate costumes of the Romans were introduced among the Germans, greater laxity in morals accompanied the innovation. The German women whom Tacitus saw "with paps visible" lost their purity when they were clothed in more costly garments. This craving for many and varied dresses among the daughters of workingmen is fatal to social progress. It devours the wages of the men, condemns many women to single life, and leads to sterility after marriage. The gown and the hat bring domestic infelicity into the homes of men who are anxious to pay their bills and lead an honest life. History teaches us that the true progress of society is not secured by appropriating the glitter and luxury of civilization. Unless wages are wisely used and the needs of society as a whole are considered, social progress will not be secured. This fact must be especially remembered by the wage earners, and it needs emphasis in society in general, for we lose sight of the old maxim that "life is more than meat and the body than raiment."

The Sclav women who come to these coal fields are susceptible to the influence of fashion. When they first come their heads are covered with silk scarfs of many colors. If she is a matron, who has passed the meridian of life, she will continue to wear the scarf; but if younger, and her pride in bodily charms is still strong, the bright-colored scarf is discarded within six months and a hat donned having a profusion of bright flowers which makes it ludicrous. The corset is also assumed, the silk waist put on, and a gown of modern fashion purchased. These articles of a higher civilization ill become the Sclav woman, whose youth was spent in farm labor, and whose form stands in striking contrast with the pinched waists and tight-laced figures of our women. The daughters of these Sclav mothers, however, wear these articles with grace equal to that of the daughters of mine employees of other nationalities. The figure of these young blondes differs widely from that of their mothers. They are not so angular; the type, according to the standard of Ranke, has been improved. The Sclav daughters come in contact with young people of other nationalities in the schools and in the social life and are reared in an environment wholly different from that of their parents. The effect is apparent in a more graceful form, a better taste in dress, and a larger expenditure of money in clothing. A Pole, in Lackawanna county, surprised a furniture dealer in the fall of 1901 by purchasing a $100 parlor suit and paying cash for it. The reason was that he had three graceful daughters in the home coming to womanhood — all native born.

When the Sclav woman buys a gown or a hat the husband invariably accompanies her and his taste decides the purchase. Many of these men are indulgent, and the custom of giving surprises to wives, so prevalent among us, is copied by them. Last Christmas, one of these men bought a silk gown to surprise his wife. It was a great surprise to the storekeeper, for the women among these men are not recipients of petty favors and delicacies such as are bestowed by indulgent native born husbands on their helpmeets. The male Sclav is lord of his house. There are exceptions. Last summer, while in an office of a justice of the peace, a Sclav was brought in by the constable charged with attempt to defraud. He was passive as many of them are. But suddenly his wife came on the scene and immediately the affair became dramatic. She argued with such vim and turned from constable to creditor and again to the justice of the peace with dramatic action worthy of a Terry or a Siddons. She saved two dollars in costs. When the storm was past, the constable said: "She's a holy terror." "Yes," added the justice of the peace, "two years ago she killed her husband by throwing the boiling contents of a coffee pot into his face, and six months after that sheep-head of a man married her." Evidently that man lived under muliocracy.

The Sclav woman, like her Anglo-Saxon sister, dresses for display. Rivalry prevails among them as among others. The hat covered with bright-colored flowers and the silk waist are worn on Sunday to church. An hour spent in watching these women as they come and go to their devotions is enough to show how vanity dominates their lives as it does the élite. Human nature is the same everywhere.

As soon as the Sclav woman crosses the threshold of her home the penurious habits of her ancestors take hold of her. The hat, waist and gown are safely stowed away, and the daily garb, scanty and unclean and torn, is put on. The shoes are cast off and barefooted she goes about her household duties.* Lippert says that many tribes cover themselves when they go out but in the homes they are naked. We have seen Sclav women in mining patches, whose scanty clothing resembled more the German females whom Tacitus saw, than women of modern civilization. Never do they wear shoes in the home. In this they are not alone. The discomfort of modern ready-made shoes is such that most people do as the Romans did — remove them when they entered the house. However, very few of the English-speaking women are found in their home without hose or slippers. The majority of the Sclavs go barefooted. Even in winter, when ice and snow cover the ground, it is nothing unusual to see these women stepping from the house barefooted to get water or fuel. Krauss says this is among them a practice to guard against colds. They go to the culm dump the very same, and some of them, though this custom fast dies away, go to church barefooted. In the home the Sclav woman attempts no decoration. There no bright ribbons, no fancy work, no cushions of unique design, no lace curtains, few shades, etc., are seen — objects in which the daughters of the Anglo-Saxons take so much pride. These women, who have to carry coal and chop wood, attend to the household duties and wash the backs of four or six men every day, have no time to spare for fancy work, providing they had the taste for it. Their life is too intense for that and fancy work costs money. That is a pastime of the disciples of Neo-Malthusianism who have multiplicity of wants. The children of these Sclavs in summer time wear very little clothing. It is nothing unusual to see them playing in the streets stark naked. Generally, however, they are covered with a calico dress and are not kept in the house or in the fenced yard in a clean white frock, and the mother ever shouting, "Keep from that dirt." They play and tumble in the dirt and they are the healthier and better to meet the hard work which awaits

* Adam Smith spoke as follows of Scotch women in his day: "In Scotland, custom has rendered them [leather shoes] a necessary of life to the lowest order of men, but not to the same order of women, who may without any discredit walk about barefooted."

them. It costs more to clothe them in winter, but even then they are more scantily clad than the children of Anglo-Saxons.

In striking contrast with all this stand the wife and children of English-speaking people in these coal fields. The children may go for two months in summer bare footed, but 50 per cent. of them are not allowed to do that. All of them wear shoes for ten months of the year and generally it means ten pairs of shoes for each child annually. In the summer, they are lightly clad and have from four to six white dresses and garments that are easily soiled; they are summer clothing. For winter they have a different supply made of heavier goods. As the children grow, the custom of two sets of clothing to suit the season of the year is kept up. The wives of native born young husbands have one or two new dresses in a year and two new hats. Their mothers were glad to get one dress in two or three years, and their hats or bonnets were often trimmed for "just one more season." "Young America," under the stress of the rising standard of living preached to them daily in the woman's page of the penny newspaper, feels that the changing seasons must be observed by a change of dress and the tendency is continually to increase the annual expenditure for clothing.

It is impossible to give accurate statistics as to the money spent by the women of the various classes in dress. It varies in every home. Among the Sclavs, the women would hardly spend \$25 annually in clothing. Among the lower class of English-speaking peoples the females would spend about the same sum. Rising to the class above this, a dressmaker of many years' experience said, the average woman would spend annually from \$50 to \$60 in dresses. There is a class still above this, which would spend annually from \$100 to \$150 in apparel.

The variation in the expenditure for clothing is not so great among the males of these coal fields as it is among the females. There is a variation, however. A clothier in Schuylkill county said: "These foreigners are peculiar. The Hun wants a suit for \$5. The Pole will rise to \$10. The Lett will pay \$15 and wants a Prince Albert coat." The difference is also seen in the shoe-store. The Hun will bring a dollar for a pair of shoes and no more will he pay. Comfort and elegance are secondary considerations. The first thought is the price. Many of these people have no hose; all they have is a piece of cloth or linen wound around the foot and leg. This, ethnologists tell us, was the practice among nations when first they found it necessary to protect the arms and legs against the rigor of a northern climate. Romans found it necessary to do so when they carried their campaigns against the peoples of the north. That was the fashion before the sleeve and the hose were invented. The first pair of stockings worn in England was by Queen Bess in the sixteenth century and, then, they were thought a present worthy of royalty and proffered by the Spanish ambassador. Stockings and sleeves, as well as linen and woolen clothing are, comparatively speaking, of modern invention, and among the Sclavs customs of antiquity are better preserved than among more civilized people. That is the case with the linen strip which serves for hose for the Pole or the Hun. The Sclav as a rule will not pay more than a dollar for a hat, while most of them can do very well without collar and tie, white shirt and an overcoat. Of course, there are exceptions. Some of the young men of these nationalities dress as well as mine employees of other nations.

The Anglo-Saxons pay for a suit of clothes from $15 to $25. Many of them wear tailor-made suits. They never go without collar and tie, cuffs and white shirt, studs, buttons, a gold watch and chain, and often a gold ring. They pay for their shoes from $2 to $3 and about the same for their hats. They never patronize a second-hand clothing store as do many of the Sclavs, and in cold weather each has a comfortable overcoat, and many of them have two, one for fall and spring and the other for winter.

The use these people get out of their garments varies greatly. The Sclav seldom parts with his coat or hat as long as the article holds together. It may fade and become threadbare, but it is not discarded. Some of the old Germans and Anglo-

Saxons show the same tendency, but not so "young America." In his case, the coat that is faded and threadbare is cast away. The hat which served last year is not worn this. The changed style in collar and tie catches his eye and his purse. He has his circle of acquaintances and he must keep up with the fashion as does his wife — if he can afford to marry. Hence, whereas the Sclav would buy one suit in two or three years, the native born buys one every year, and his supply of linen and underwear is an item of expense which little troubles the Sclav. Here again we cannot give accurate statistics. But the average young man of native birth, married or single, would spend from $40 to $50 annually in clothing; the Sclav would not spend one half that.

A banker in a mining town said lately: "The only people who save money to-day in the coal fields are the foreigners." The reason is apparent. It will become still more apparent, in the study of the variation in the standard of living as seen in food and amusements.

The Money Spent on the Table and on Amusements.

Lippert tells us that the Indians of New England called the dwellers of the north "Roheesser." As men began to cook their food they were called the "cooking men." The same authority says that man's advancement in civilization can be measured by the command he has over the supply of food within reach of him. In these two directions men advanced: on the one hand, we have the art of cooking, and on the other, we have a greater variety of food for the maintenance of human life.

The art of cooking is, comparatively speaking, of recent invention. Buchanan says that the Irish in the sixteenth century had no tables from which to eat their meat. They placed it on bundles of straw. The meat was cooked by selecting a hollow tree, placing the meat therein, and then setting the tree on fire. A fallen horse often furnished them meat for the feast. The Celts down to the seventeenth century warmed their milk by putting into it a heated stone. Even to-day, many so-called civilized people prefer meat raw than cooked, and it is not an unusual sight to see Sclavs eating a piece of raw pork. And, notwithstanding our advancement in the art of cooking, Robert Blatchford says that the wives of workingmen in "Merry England" do not know how to cook a beefsteak. The wives in our towns fall more and more into the habit of buying bread for family use or subsist on hot biscuits.

The variety of food found on the table of the average workingman is surprising. All the ends of the earth contribute to the needs of the wage earner to-day, and if his supply of one article of consumption runs short, he has a large number of others to fall back upon. The laborers of our country are fortified against hunger by a series of commodities which stand in a gradually ascending scale of refinement, so that, in time of hardship, they can fall back upon grades of cheaper food and be better able to subsist and carry on the conflict of life than former generations were. Among wage earners to-day, the "felt want" as economists say, is far above the "real want," and in the strike of 1902, the mine workers of the anthracitc coal regions curtailed their purchases to such an extent that the merchants who dealt in articles of prime necessity, such as meat, butter and cheese, did not sell half the amount they sold when the mines were operated. A potato famine in Ireland a century ago killed thousands of the inhabitants; if wheat were wholly cut off to-day from the tables of American workmen no such result would follow.

No country on the face of the globe has a greater supply of food than America. General Booth, of the Salvation Army, said, after one of his visits to our country, that the food wasted in our homes would feed a nation. Isola Deschenes, writing on the extravagance of American families, says: "Cut down the size of the garbage barrel — my observation has taught me that about one third of the uncooked food that goes into the average house goes out in that receptacle for refuse." An economist said: "Of two or three hundred weight of provisions, which may sometimes be served up at a great festival, one half perhaps, is thrown to the dung-hill, and there is always a great deal wasted and abused." The same may be said of our festivities. Even in church festivities — among the followers of the Nazarene who said "gather the fragments," the waste of the gifts of nature is shameful. It all comes from the superabundance of food at our command, which is supplied in great variety to the workers of our country.

European workers who immigrate here feel this. In a feast given in honor of one of our coal operators of Lackawanna county, the gentleman who made the speech of the evening said: "We often hear men speaking of the roast beef of old England, how much better it tasted than ours does. No wonder, they only got a taste once a day, while we eat it three times every day." Adam Smith said of the French and Scotch laboring classes of his day, that they "seldom eat butcher's meat, except upon holidays and other extraordinary occasions." That was exactly the condition of most of these Sclav immigrants in the fatherland. There they seldom eat meat; here they get a plentiful supply at reasonable prices. Is it then surprising that most of them think this a goodly land? Nothing proves the economic advantages of our country better than the richly laden tables of our working classes, and those of the employees in the anthracite mining regions are no exception.

The thrifty Sclav immigrant soon finds that the traditional standard relative to the quality and quantity of food consumed by his countrymen is materially changed here. When these men first came into the country, it was nothing unusual to see a company of 20 or 30 men leading a communal life in a large barn. The place was run by a boarding boss and his wife. Each man paid a dollar a month for sleeping room, and a little extra to the woman for washing his back each evening. Meat, potatoes, coffee, bread and cabbage were bought in common. At the close of the month, each paid his pro rata share, which was about $5. One of these men said, if his share went up to $6, "Me kick": the cry of extravagance was raised and there was war in the camp. A change has come. Now single men pay from $2 to $3 a month for lodging, washing, etc., and buy their own provisions. It costs them under this system about $10 a month. Many Sclav young men in recent years, following the American fashion, board and pay $12 a month. The Anglo-Saxon boarders pay from $16 to $18. This amount was not paid by immigrants from the British Isles in the fifties and sixties of the last century. Young men then boarded in Carbondale and Minersville for $10 a month. The wages at that time were $1 a day for miners and 75 cents for laborers. But food was cheap. A quarter of beef could be got for 3 cents a pound; a whole sheep for $1 or $1.25; potatoes for 25 cents a bushel; butter was a shilling a pound; eggs 8 cents a dozen and flour $5 a barrel. Anglo-Saxons also lived then simpler than they do to-day, and their "felt want" was much closer to their "real want" than it is now. In this condition the majority of Sclavs find themselves at present.

A study of the day-book of stores, where Anglo-Saxons and Sclavs deal, reveals very clearly the difference in their standards of living. A greater variety of articles are consumed by the former than the latter. A store-keeper said, if the bill of a Sclav goes up to $10 a month for groceries, it is high; the bill of the average English-speaking family goes up to $20 and $25. By a computation made, in one of the company stores in Schuylkill county, of the purchases of 12 English-speaking and 12 Sclav families for one year, we found the per capita expenditure of the former to be $5.48 and of the latter $2.86 per month. In the account of the Sclavs we found the following items: flour, barley, salt-pork, potatoes, cabbage, pickles (barrel), garlic, coffee, and coffee essence, sardines (5 cans for 25 cents), eggs, and very sparingly butter, cheese, and sugar. In the list of Anglo-Saxons there were flour, ham, onions, potatoes, cabbages, pickles (bottled), coffee, tea, eggs, lard, dried beef, spices, cakes, crackers, mackerel, canned tomatoes, canned peaches, canned apricots, canned cherries, soap, rubbers, brooms, lemons, salmon, and large quantities of butter, cheese and sugar. A perusal of the contents of these books clearly showed that the felt want of the "white men" was far larger than that of the "foreigners." The dawn of luxury, however, was visible in some of the Sclav accounts. It appeared in the purchase

Sclavs Selling Huckleberries to the Shipper.

of cheap prunes, mixed jams (5 pounds for 25 cents), and a brand of apple-butter (3 pounds for 10 cents). These luxuries would go a long way. Observing merchants say that a Sclav family will live on half what is thought to be necessary for the maintenance of an equal number in an English-speaking family.

Of course, there are exceptions. We knew an Anglo-Saxon who divided a herring for two meals, thinking it luxury to eat the whole of it at once. No Sclav can surpass that save the fellows who make a meal on bread alone. Among the English-speaking of these coal fields, the Germans have the credit of practicing greatest economy in the home. As above stated the contents of the tables of 50 per cent. of mine employees reflect the condition of the mining industry. A wag said, as he smelt the stench of smoked herrings in a patch: "It's poor times; when times are good you'll smell beefsteak and onions." He spoke the truth. When the pinch comes the table feels it even sooner than dress or social amusements. Vanity, even in civilized people, is stronger than appetite. Many a girl goes ungrudgingly to the table to satisfy her hunger with bread and pickles if only she can get that waist made for the party. Many a family also, which struggles to meet its dues in a building and loan association, will spend less on the table than those who have their homes paid for. It is wonderful how some will stint themselves for the sake of a house. It is pathetic to hear a father on his dying bed saying: "How foolish, why did we pinch ourselves so?."

The Sclavs as a rule make the best of everything. If they have a garden, they take good care of it. Many of the Anglo-Saxons do the same. It is interesting to pass along the Schuylkill and Tremont valleys and see the many little farms which are cultivated by mine employees of the Philadelphia and Reading Coal and Iron Company. In the strike of 1902, hundreds of mine employees' families could not have carried on so brave a fight if it were not for the small farms and large gardens they cultivate, which are leased to them or are attached to the company houses.

In the summer months, the wives and children of the Sclavs gather huckleberries on the mountains, which they sell from house to house, or to hucksters who crate and forward them to the city. Few Anglo-Saxon women pick berries; they say: "What's the use, when you can buy them at the door so cheap?"

In amusements the English-speaking spend far more than do the Sclavs. The continual series of theatrical companies which come to our towns are almost wholly supported by the native born. To the Sclav these are no attractions for they do not understand the proceedings. His chief diversion is the saloon, card-playing, an occasional dance and the weddings and christenings which occur. Those of the baser sort carouse after each pay, and when their animal nature is excited and the brute appears, they become more fierce and blood-thirsty than the savage beast of the forest.

The mother of a young American said: "He takes $10 every month and spends it, and on Thanksgiving and Christmas asks for extra." A young Sclav will live on that. Young men, earning their living in and around the mines, spend each month on amusements from $3 to $5. In every town of from 4,000 to 6,000 population there is an opera house where theatrical companies play, and it is estimated that from $6,000 to $8,000 is spent annually by the native born in amusements. In a town of 13,000, the amount annually spent was estimated to be from $20,000 to $25,000. All this comes from "young America," and it partly accounts for the fact that this latter class saves no money, while the young Sclav lays aside each month from $20 to $25 when the mines are working regularly.

Much is said about the food supply of the Sclav, and it is claimed that it is not what it ought to be. The question, what amount of food does an adult need to replace the force spent in labor and keep intact the physical organism, is variously answered. Charles Richet, a learned French physiologist, gives the following amount of food as adequate to satisfy an adult for 24 hours: Meat, 4.4 ozs.; bread, 10.5 ozs.; potatoes, 10.5 ozs.; butter and cheese, 1.7 ozs. He said that we daily exceed this limit to the injury of our health. The adult who would limit himself to the above supply would live on an average of about 13 cents a day, or $3.90 a month. The Hun's expenditure exceeds that not counting the beer he drinks and the few luxuries he indulges in.

Dr. Ranke gives the following quantity of food as sufficient for 24 hours for an adult who performs arduous toil.

	Grammes.		Ounces.
Proteids	110–120	or	3.8–4.2
Fats	60–100	or	2.1–3.5
Carbohydrates	450–500	or	15.8–17.5

Professor Huxley gives the following quantity as "an average daily diet for a healthy man."

	Grammes.		Ounces.
Proteids	130	or	4.6
Fats	50	or	1.7
Carbohydrates	400	or	14.1

To yield this quantity of food-stuff all that is necessary is

	Ounces.		Cents.
Very lean meat	8	worth	7
Bread	16	"	7
Potatoes	24	"	1.5
Milk	00.75pt.	"	3
Fats	1	"	1
Total			19.5

According to this estimate a healthy man's diet would only cost him 19.5 cents a day, or $5.85 a month, which approaches nearer the Sclav's expenditure than that of Richet.

The lowest stratum of mine employees has a larger variety of food than the list laid down by Richet or Huxley. The Sclavs have good bread made of the best wheat or rye; they use much barley in soups; they consume daily about a pound of fat pork or beef for boiling (8 to 10 cents per pound) or bologna, sausage, a quantity of potatoes, cabbage, milk, coffee and beer, butter and cheese, sugar, garlic, and on fast days eggs and fish.

Some may object to the Sclav's way of eating — a plain table, agate utensils and a pocket knife. A table-cloth of fine linen, imported crockery, silver knives and forks, etc., certainly aid the appetite, but if according to Richet we impair our health by eating too much, why intensify the evil? But fashion? Yes, that is the nemesis — the tyranny of fashion; inexorable rivalry among the working classes, who imitate in a shallow manner the rich in house, clothing and diet.

What is the standard of living for American citizens of which so much is said of late? Does it mean meat three times a day, a quart of beer and an ounce of tobacco? Or shall we take that of homes with a large garbage barrel into which one third of the food purchased goes? Or shall we say that the American citizen ought to dine at seven, spend his evening in the theatre, then play poker till midnight and go to bed "tight." This indefinite term the "American standard of living" affords the demagogue a lever to arouse an audience of wage earners to enthusiasm, but the sooner society, and especially the working classes, learn that no organization or legislative measures can secure them a life of ease and luxury, the better. We can only get from nature our food, clothing and shelter by arduous toil, and anyone who teaches otherwise will lead the working classes into the slough of despair whence they cannot extricate themselves save by hardships and bloodshed. A good appetite, waiting on plain diet, is far better than a sluggish digestion deranged by highly seasoned food. Mephistopheles tells Faust he can at eighty grow hale and young, if only

> ". . . to the fields repair,
> Begin to delve, to cultivate the ground,"

against this the scholar and gentleman protests:

> "For this mean life my spirit soars too high,"

and to the witch he goes, where for a cordial he sells himself to the devil and henceforth walks the way where lust, murder and hell await him.

This same Faust-like spirit is still in the world. Nitti says that the cry for higher wages and fewer hours is the outcrop of repugnance to manual labor. Men dream that by some magic spell they can secure themselves the blessings of earth which only come by honest toil. There is only one outlook for this. It is the way of sorrow, hardships, anarchy and murder. Dr. Ranke says that every man is healthier and better if his physical powers are exercised within the range of their capacity, and the repugnance to manual labor, so commonly seen among the descendants of foreign born parents, cannot be regarded otherwise than as a degeneracy.

The Effects of a Rising Standard of Living.

We have seen the great variation in the standard of living between the Sclav and the Anglo-Saxon. The peoples of southern Europe come here with their traditional customs and usages. They cling to them with great tenacity, but it would be a mistake to suppose that they are not influenced by the nations around them. The Sclav, notwithstanding he is behind the Anglo-Saxon in civilization, is not so far removed that he does not feel the influence of the law of social capillarity. Walter Bagehot said that "the experience of the English in India shows that a highly civilized race may fail in producing a rapidly excellent effect on a less civilized race, because it is too good and too different. The higher being is not and cannot be a model for the lower; he could not mould himself on it if he would, and would not if he could. But in early society there were no such great differences, and the rather superior conqueror must have easily improved the rather inferior conquered." The relation between the various nationalities in these coal fields is best represented by that existing among men in the early stage of the world's civilization. They act and react upon each other, and the Sclavs gradually feel the effect of new ideas in a new environment.

Among the representatives of the races from the British Isles there has been spirited rivalry for the last half century. It is even now felt by the descendants of the various races. This, as shown in the last chapter, enters into the economic life of the male members of our population. In the sphere of the home it is none the less apparent among the female members of our society. This rivalry brings with it many evils, but it is also productive of great good. Lilienfeld says that a society without conflicting interests, competition in production and conflict for existence, is doomed to death. A stratified society is the curse of India. That is not the case with our society. The law of social capillarity was never felt by man as it is in our land, and the anthracite coal fields are no exception. Into this competition and conflict the Sclavs are entering more and more, and the tension and stress of life felt in these communities in the past fifty years are destined to characterize still further them for the next half century. Mr. Ward tells us that "in the last analysis the sole fact in the idea of life" is irritability. In recent years the Sclavs have come into closer contact with Anglo-Saxons than at any former period since their introduction into the coal fields. One of the miners said: "The Sclav hitherto has been led, soon he will lead." The irritability introduced into his life since 1900 has awakened his energies as never before, and the activity and leadership displayed by many of them foreshadow a competition between Sclav and Anglo-Saxon in this area such as was never waged between the Teuton and Celt.

The change wrought in the standard of living accounts for the industrial friction which has seriously disturbed the anthracite industry of recent years. Industrial conflicts were fiercely waged in these coal fields in the latter seventies of the last century, but the causes were more due to the fluctuations in the monetary affairs of the nation than to any other factor. In the strikes of 1900 and 1902, the chief cause of the conflicts is found in the changed economic life of a large number of mine employees. No great movement can be adequately explained by one cause. A composition of forces ever act upon society as in the inorganic world, but as in the latter the direction of the moving body indicates which is the prime motor in the combination, so the emphasis laid upon the economic life of mine workers indicates that the chief cause of murmur lies there. In his public addresses, John Mitchell, President of the United Mine Workers, always emphasized the fact that the wages paid the employees of the anthracite collieries were not adequate to afford the men a living such as American citizens ought to secure. He spoke as a native born citizen, and he mouthed the sentiments of the thousands of native born employees in and around the mines, who regulate their living according to a standard much higher than that by which their foreign born parents lived.

The wages of laborers determine the amount of articles of consumption placed at their command. The real wage is to be measured by the quantity of consumable goods which the wage earners can purchase with their earnings. Adam Smith said that the better wages paid labor in England than in France accounted for "the difference between the dress and countenance of the common people in the one country and in the other." It is also true that a "rise in the average price of necessaries, unless it is compensated by a proportionable rise in the wages of labor," must affect the social and domestic relations of the wage earners. Down to a certain point, vanity will cut across the line of real need. Put the marginal line of necessaries ever so low, when that is reached, men will part with all superfluities that they may keep the organism in working order. These laws, so well established in economics, should be laid along side that of social capillarity in order to explain the conditions in anthracite communities which precipitated recent conflicts between capital and labor.

Dugald Stuart said: "Far from considering poverty as an advantage to the state, their [modern politicians] great aim is to open new sources of national opulence, and to animate the activity of all classes of the people, by a taste for the comforts and accommodations of life." The same idea is frequently met with in modern writers. C. R. Henderson says: "A civilized man wants many things and is willing to work hard to get them." This doctrine accounts for the fact that modern industry has so stimulated production that the markets are flooded with new commodities which, entering into the lives of men, are soon deemed necessaries. Lord Chief Justice Hale computed in the time of Charles II. that "the necessary expense of a laborer's family, consisting of six persons, the father and mother, two children able to do something, and two not able, at ten shillings a week or twenty six pounds a year." Nitti puts the annual expense of the average family of mine workers in the countries specified as follows :

United States	$655.88
England	571.85
Belgium	466.70
Germany	461.73

This is six-fold what it was in the time of Chief Justice Hale. Of course the prices of provisions have greatly advanced since the seventeenth century, but the chief difference lies in the rise in the standard of living among the working classes. Indeed the real need of mine employees in the United States is far higher than that of their brethren in Belgium or Germany, as the differences in the above averages suggest. This rise in the standard of living is stimulated by no class as by capitalists, and, if the wage earners demand higher wages to meet the larger list of commodities now deemed necessaries, they ought not to complain of the "insolence of employees." If capitalists, having goods to sell, stimulate the wants of consumers and designedly create tastes among the masses for articles of neutral and even negative utility, they can expect a harvest of industrial conflicts which will, sooner or later, lead to experiments in some form of socialism. The law of nature is "ubi stimulus, ibi fluxus." This is also a law in social life. Professor Clark has shown that at present a larger proportion than ever before of the productive wealth goes to laborers. It is also true that interest decreases as capital increases. Granting this to be true, it is yet equally so that the fluxus does not keep pace with the stimulus in the working classes, and the result is that laborers, under the sting of a rising civilization measured by the multiplicity of our wants, are up in arms against the captains of industry, and demand higher wages to purchase more consumable goods and shorter hours to enjoy them. The Anglo-Saxon has taught the Sclav the new demands and stubbornly does the latter stand by his leaders in the conflict.

What is to be the criterion of the necessaries of life? At the end of the fourteenth century an English writer condemned the refinement of his contemporaries in constructing chimneys instead of leaving the smoke pass where it would. He also condemned, as uncalled-for luxury, the substitution of wooden vessels in place of clay ones. He did not think that chimneys and wooden vessels were necessaries. A very different standard is given us by Mr. Ward who says: "Everything which exerts the least influence in improving the physical condition of man is a necessity." A bath, clean linen, comb and razor would greatly improve the physical condition of fifty per cent. of the mine employees, but they are far from considering these conveniences — accessible to all — as necessaries. The same author says: "It is not only necessary to maintain a bare existence, but it is essential to the preservation of the race that its means of subsistence be ample and abundant." If we consider this from the standpoint of the variation in the standard of living between Sclav and Teuton, as shown in the previous pages, we see that what affords a bare existence to the latter is ample and abundant to the former. While the Anglo-Saxons are complaining that they are not able to make a living in these coal fields according to the wages paid in 1901, the Sclavs save money. A Hun, who settled in Lackawanna county in 1892, worked for $1.38 a day and his wife kept boarders; in 1900, both returned to the fatherland with $2,000 saved. A young Ruthenian began life in the coal fields in 1890 as a laborer and in ten years had saved $1,500. Private bankers, who have kept these people's money, say that the average man among them will save from $20 to $25 each month when the mines work regularly. Here are some examples. A young Sclav, who had been in the country nine months, had the following bank account: March $55, April $20, May $20, June $20, July $44; making a total of $159. Another, in the country two years, had the following bank account: 1899, July $30, August $20, September $20, October $25, November $30; 1900, January $25, March $20, May $25, June $20, July $25, making a total of $240. These are not exceptions.

They are but fair samples of the thrift of this people. As a rule they save money. It is otherwise with the Anglo-Saxons. New wants and desires have multiplied in the life of the native born so that he spends, in luxury, a monthly sum amply sufficient to procure the Sclav the means of subsistence.

Adam Smith says: "The principle which prompts to save is the desire of bettering our condition, a desire which, though generally calm and dispassionate, comes with us from the womb and never leaves us until we go into the grave. In the whole interval which separates those two moments, there is scarce perhaps, a single instance in which any man is so perfectly and completely satisfied with his situation as to be without any wish of alteration or improvement of any kind." This was preeminently true of the immigrants into these coal fields from the British Isles and Germany the past generation as it is today of the Sclav. And, not only were these men anxious to improve their lot, but they were willing to work hard, live on plain food and clothing, and practice frugality. The descendants of the earlier settlers are not so industrious, not so frugal and not so simple in their tastes as their fathers were, so that while their desires and wants are higher their productive capacity in the mining industry is lower. The maxim, "every prodigal appears to be a public enemy, and every frugal man a public benefactor," is largely forgotten now-a-days. Men are anxious to make work. The native born will drive his pick through the empty powder-keg which his father saved for the sake of the 10 cents it would bring. One of the local unions in our towns passed a resolution, that no man should patch old rubber boots for, by so doing, he diminished the labor needed in that industry. Prodigality characterizes "young America" in these regions, and not only does he beggar himself thereby, but the country is also impoverished.

In the last strike (1902), the president of the Miners' Union asked that two church dignitaries be appointed arbitrators to decide whether or not wages paid anthracite mine employees were adequate to give them a living such as American citizens ought to have. Suppose half a dozen farmers from the valleys surrounding the coal fields were selected arbitrators in the dispute? Would the man with the hoe say that the man with the pick did not get his share of the productive wealth? How much is the annual income of the farmer—this man who has during the years of tariff tinkering and manufacturers' greed and industrial selfishness, patiently borne the burden until many of his ilk in the Eastern and Middle States have been driven to bankruptcy and despair! When will the eyes and hearts of wage earners see and feel that this fatal idea of "making work" and the Utopian idea of "living like gentlemen" only add to the burden that must be borne by their brethren either on the soil or in the lower stratum of workers in the industrial world? Labels of trades' unions and of consumers' leagues mean advanced prices for commodities purchased by the working classes themselves. It is a tax placed by brother upon brother. The selfishness of some trades' unions has never been equaled by the most tyrannous of capitalistic monopolies. Upon whom fell the suffering and loss of the anthracite coal strike of 1902? The impoverished mothers and ill-clad children of the coal fields, the wage earner's family and the sickly poor in the cities; these were the ones who suffered the brunt of that conflict. And it is so in every industrial strife. One class of wage earners in the attempt to raise its wages inflicts suffering and death upon another class of wage earners. Such crude methods of improving the lot of man is not worthy of the intelligence and the civilization of the twentieth century.

Is there a better way? Not so long as workingmen, driven by the desire to improve their lot, base their hopes upon objective realities, while neglecting the man—the center of our civilization. Shoddy goods, glittering trinkets, cheap upholstery, and legislative quackery, will never improve the lot of man. The Phœnician, Grecian and Etrurian traders, who frequented the trading stations of the Volga and Dvina, presented a great contrast to the barbarous nations of northern Europe. The civilized, who were refined and had a great multiplicity of wants, brought glimmering metals and shining glassware in exchange for the skins, oil and cereals of the barbarians. These last commodities were useful, and in time their producers conquered the producers of articles of neutral and negative utility. The simple barbarians on the shores of America were surprised at the passion of the Spaniards to obtain their gold ornaments. These glittering baubles they gave to their new guests without seeming to think they had made them any very valuable present. But that thirst for gold proved the death-knell of the nation. An insatiable thirst for material things will doom every nation to degeneracy.

In the evolution of society, success in the conflict of life depends more and more upon the psychical factors of the contestants. In primitive, and indeed down to comparatively recent times, the physical sphere of the conflict was all important. In recent times, this is changed. It may be true as Mr. Ward says, that in the last analysis all conflict is reduced to force, but the change in the character of force is all important. In the conflict between individuals, societies, or states, intellectual capacities and moral qualities decide the victory. Purely animal tendencies count for little now-a-days and, in the future, the psychical sphere of the conflict will still grow in importance.

This great truth should be learned by the workers who try to improve their condition. The virtues of frugality, simplicity and thrift cannot be abandoned. The supply of wealth is limited, and any attempt to secure unto all men the luxury and abundance of the rich must end in misery and suffering. Goethe once said : "It is always a misfortune for him [man] when he is induced to strive after something with which he cannot come into active relations." The wants of the native born in these coal fields are so stimulated that they cannot be satisfied by the productive wealth of these mines. As a consequence "young America" is up in arms against the social and industrial order which does not supply it with the means requisite to maintain this raised standard of living. Sooner or later society will find out that this is not the path of peace. It seems as if "whirlwinds of rebellion" must first "shake the world" before the lesson is taught, and perchance the pessimism of Galton and Rumelin, that a return to barbarism is the only outlook before modern society, is not the forecast of timid hearts.

If such a doom is to be evaded, the working class must give more attention to the psychical factors. An English writer of the sixteenth century complained that oak was used in the construction of homes which formerly were built of willow, and adds : "Formerly the houses were of willows but the men of oak, but to-day it is the contrary." The progress of society must ever rest upon the type of man at the center of it, and in our studies we always come to this one result, that the only hope of man's advancement lies in the physical, intellectual and moral improvement of the individual. *

*The comparison of Sclav and Anglo-Saxon made in this chapter is not exactly just to the Sclav, for the English-speaking mine employees have been under the influence of American civilization for over a generation. It would be more just to make the comparison with the immigrants of thirty or forty years ago from the British Isles. The late Abram S. Hewitt spoke of a tour he made through the coal fields in 1876 as follows: "I found terrible conditions there. I found the men living like pigs and dogs, under wretchedly brutal conditions." Nothing worse than that can be said of the Sclavs of to-day.

CHAPTER V.

MINE EMPLOYEES AT HOME.

1. Dwelling Houses in Mining Towns. 2. The Condition of Women Therein. 3. The Rights of Children. 4. The Need of Better Homes.

Dwelling Houses in Mining Towns.

According to the census of 1900, the average number of persons to a family in the State of Pennsylvania was 4.8, and the average number of persons to a dwelling was 5.1 in 1900, 5.3 in 1890, and 5.5 in 1880. In the counties of Lackawanna, Luzerne and Schuylkill, 5 persons is the average number in a family and 5.4 to a dwelling. Thus we have a slightly larger average in both these respects in the counties where anthracite mining is the staple industry than in all the State.

The following table gives the population, the number of dwellings and the number of families in nine counties of our State.

County.	Total Population.	Total Dwellings.	Persons per Dwelling.	Total Families.	Persons per Family.
Lackawanna	193,831	34,454	5.6	38,779	5:0
Luzerne	257,121	48,070	5.3	50,226	5.1
Schuylkill	171,927	33,236	5.2	34,328	5.0
Berks	159,615	33,173	4.8	34,594	4.6
Susquehanna	40,043	8,976	4.4	9,592	4.1
Chester	95,695	20,206	4.7	20,768	4.6
Lancaster	159,241	34,323	4.6	35,559	4.5
Clearfield	80,614	15,467	5.2	15,879	5.1
Westmoreland	160,175	30,111	5.3	31,738	5.0

In this table the three first counties are in the anthracite coal fields, while the two last are in the bituminous coal fields, and the intervening four counties are chiefly agricultural. The anthracite and bituminous counties are practically alike in the

An Eviction After the Strike of 1902.

number of persons to the family and to the dwelling, while they lead the agricultural counties in both respects.

If we carry out the comparison to purely mining towns the contrast is still more striking.

	Place.	Total Population.	Total Dwellings.	Persons to a Dwelling.	Total Families.	Persons to Family.
Anthracite.	Mahanoy City	13,504	2,529	5.3	2,584	5.2
	Shenandoah	20,321	3,521	5.7	3,683	5.5
	Olyphant	6,180	945	6.5	1,130	5.3
	Nanticoke	12,116	2,233	5.4	2,325	5.2
	Winton	3,425	572	5.9	604	5.6
Bit.	Clearfield	5,081	974	5.2	1,005	5.1
	Chester City	33,988	6,677	5.1	6,908	4.9
	Connellsville	7,160	1,367	5.2	1,477	4.8
Agr.	Towanda	4,663	1,034	4.5	1,138	4.1
	Tyrone	5,847	1,280	4.5	1,312	4.4

The table shows that in mining towns the number of persons to the family and the number of persons to the dwelling exceed those of rural communities.

If we take the five first towns specified above, we have 5.7 as the average number of persons to a dwelling in typical anthracite mining towns. This necessitates 70,175 dwellings to shelter the 400,000 persons directly dependent on this mining industry. Beside this there are other 50,000 dwellings in the territory under consideration, where dwell the professional men, merchants, mechanics, etc., who are indirectly dependent upon the anthracite mining industry.

Many of the houses in which the mine workers live are owned by the coal companies. The majority of these was built by individual operators in the fifties and sixties, when operations were begun in a wilderness and the mining companies in order to shelter their employees within reasonable distance of the mines, had to build houses for them. When the individual operators were bought out by the corporations, the houses were sold with the plant, and a large corporation, such as the Philadelphia and Reading Coal and Iron Company, continues to rent these houses to its employees.

The following table, compiled from the assessors' books of

the various counties, will give some idea of the number of houses, together with their assessed valuation, rented by the various coal companies to their employees.

COMPANY HOUSES. ASSESSORS' VALUATION.

County.	Total No.	$25	$50	$75	$100	$125	$150	$175	$200	$225	$250	$300	$350	$400	$500	$600	$700	$1000
Schuylkill	3233	98	79	99	1923	66	272	8	237	82	40	185	37	42	27	7	1	30
Luzerne	3973	20	299	97	1042		96				795		911		472		197	45
Lackawanna	731	22	11	48	121	196		159	70			84		12				8
Northumberland	605	130		50	39	90	2			44	91	56			102			
Total	8547	270	389	294	3125	352	370	167	307	126	926	125	948	54	601	7	197	83

This gives an idea to what extent the coal companies are real estate agents renting dwellings to their employees. The list is not complete; there are possibly a few hundred dwellings owned by coal operators which have not been tabulated. About 16 per cent.* of the houses in which the mine workers live is owned by the companies for whom the men work. The rent for the house is kept each month at the office, and if the company sees fit it can not only discharge the man from his work, but also evict him, which in many instances means departure from the town or village, every inch of which is owned by the company. This was done in several instances after the last strike on the Hazleton mountain, and the evicted had to seek other shelter while their furniture was piled by the sheriff on the public highway.

All sections of the anthracite coal fields are not equally under the sovereignty of the coal operators as to houses and land. In the Northern coal field employers of labor have not hemmed in their employees on all sides so as to preclude any attempt on their part to possess real estate. All the houses

* In the Report of the Anthracite Coal Strike Commission, page 43, we read: "So far as could be ascertained, the facts show that in the Northern and Southern coal fields less than 10 per cent. of the employees rent their houses from the employing companies, while in the Middle coal fields a little less than 35 per cent. of employees so rent their houses." This is not accurate. Company houses are not so numerous in the Northern as in the Southern and Middle coal fields. In the Middle coal field the percentage of employees living in company houses is not 35. In the Fifth District only do we find the percentage of employees living in company houses large and there it is about 45 per cent.

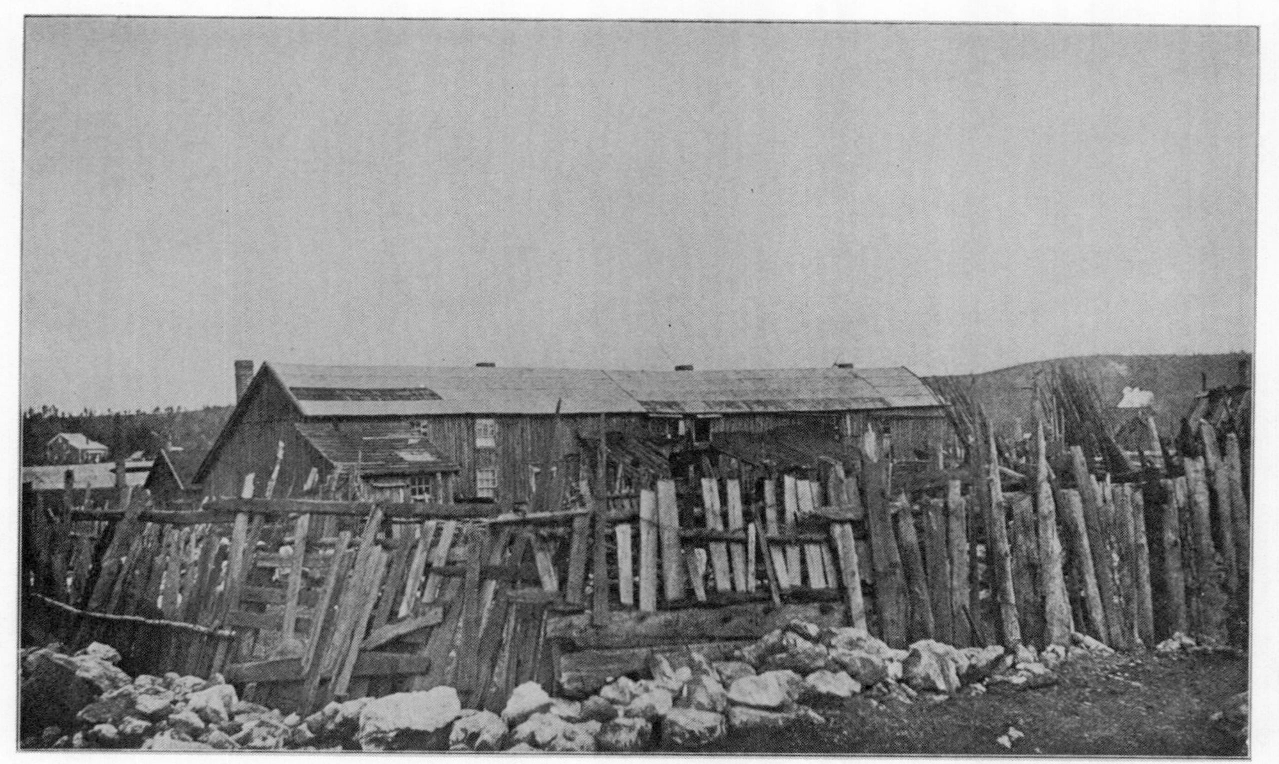

One of the Worst Company Houses in the Coal Fields. (Four families in this building.)

rented by companies in the Lackawanna and Wyoming valleys would not exceed 1,500, or about 3 per cent. of the houses occupied by the mine employees in this territory. These coal companies have sold building lots upon which houses are built and the wage earners feel that they have something to live for besides digging coal. A sense of responsibility also comes with the possession of real estate, and the man who builds a house, plants trees and cultivates a garden is a better employee and citizen. The formation of the coal basin in the Wyoming and Lackawanna valleys has something to do with the freedom given there to individual enterprise. The basin is one continuous whole, and towns have grown on every part of it, so that to-day from Forest City on the extreme north to Shickshinny on the extreme south, a continuous series of towns connected by steam and trolley railroads are found. In the Middle and Southern coal fields it is different. Here many of the coal basins are small and scattered. This favors isolation, and a company possessing one of these basins has a property which may be wholly separated from all neighboring towns. Upon this basin, owned entirely by the coal operator, a town grows which stands alone in the mountains, far removed from any town or city where freedom is given to individual enterprise. The best illustration of this is the towns which have been planted on the Hazleton mountain where the independent operators thrive. The map on page 244 shows the number of small towns owned by the coal operators clustering around Hazleton and Freeland which virtually are the only two spots on the mountain where individual freedom and enterprise are given free play. In all the other mining camps, with very rare exceptions, every inch of the ground and every house is owned by the independent companies. In these places also the company stores flourish most vigorously, and, notwithstanding legal attempts to abolish them, they still exist. The land monopoly possessed by the company gives them a great advantage, for no one from without can enter these mining towns and secure a foothold on which he can offer the necessaries of life for sale to the inhabitants.

In Schuylkill county there are many localities in precisely the same condition, but the evil does not exist to so great an extent as in the Fifth District. Within a radius of three miles of Mahanoy City there are fifteen small mining villages where all the houses are rented by the company, and in four of them the company store is also present. In mining camps outside the boroughs of Shamokin, Mt. Carmel, Shenandoah, Ashland, Minersville and Pottsville the same is true. The superintendent of public schools in Mahanoy township speaks as follows: "While we have nearly 2,000 taxables and more than 1,200 voters within our borders, we have not a single freeholder in the township, ninety-eight per cent. of the male working population being employed in or about the mines and hence constantly changing." What was there to prevent constant changes? A person who can acquire no property can have no other interest but to work as little and consume as much as possible. What inducement is there to these men to stay when no opportunity is offered them to raise their status in society? There are no interests which these men can pursue outside the daily routine of the mines, and there is no opportunity given them to invest whatever money they might save. Healthful competition between neighbors is precluded, for none will take interest in a house or garden from which he may be evicted at the good pleasure of the company which furnishes him work whereby he and his family subsist. Individual possession has ever reacted upon the birth-rate, but in these towns the tenants are not given the opportunity to feel the wholesome restraint exerted upon passion and emotion by personal possession. It seems incredible that men of keen insight would so forget the dictates of ordinary prudence as to shut off their employees from opportunities of self-improvement or self-advancement which would be their safest guarantee of good workmanship and moral conduct. This system of capitalistic greed accounts for the ease with which the population of certain localities has changed in recent years. There was nothing to bind the workers to the soil and, being forced to leave when their demands for personal and economic rights gave offense to the operators, their places were taken by a lower grade of labor, which to-day largely occupies the homes possessed by coal operators.

The quality of the company houses varies greatly. The two accompanying pictures give the extremes. In Luzerne, Schuylkill and Northumberland counties, the assessors classify many of these miners' dwellings as "shanties" and assess them at from $10 to $25. For these the companies charge from $1.75 to $3 per month. No repairing is done to them and it has been the policy of the Philadelphia and Reading Company to destroy these shells when they are vacated. But the vacating seldom happens and the company keeps on collecting the rent. Shanties have only one story and a garret, and sometimes the first floor has only one room. An additional small shed is frequently built into which the cooking stove may be removed in summer.

Over 50 per cent. of the company houses are assessed from $10 to $100. These are poor dwellings and ill afford the shelter needed by the tenants in the cold of winter. They are built of hemlock boards with weather strips nailed over the crevices. No plastering, no ceiling and no wall paper are furnished. The best of them contain two rooms on the first floor and one on the second. They are generally tenanted by Sclavs. Many of these are veritable shells of two rooms 16 x 16 feet for which the tenant pays $4 a month. One of the men told us: "Me cannot keep warm in winter," and if the members of that company were not rendered impervious by greed to the demands of humanity they would not ask these men to dwell in these shells during the winter season. This company had in the patch thirty such houses for which it charged from $4 to $6 a month. By a liberal computation, the cost of erection would not exceed $400 or $500 per double house, so that the property yielded from 25 per cent. to 30 per cent. annually. A miner who had lived in one of this company's houses for thirteen years said that during that time no repairs had been done by the company. He had to do them himself.

In the care of company houses no rule can be laid down. The same company deals differently with the different grades of houses in its possession. The Philadelphia and Reading, the Lehigh and Wilkesbarre and the Cross Creek Coal Companies have many dwellings which are classified as shanties and which rent for from $1.75 to $2 a month. These are not fit habitations for men. The companies admit they do not spend anything in repairs on them. If the tenants wish to live in them and do what repairs they may so as to make them tenantable, they may do so ; if not they may vacate them and the shanties are torn down. Torn down they ought to be in any case, for the drunkard or the curmudgeon ought not to have the opportunity to pen his family in a miserable hut in order that he may spend more in drink or save the dollar which should be spent in securing proper shelter for the family.

The following list of houses rented by Coxe Bros. & Co. gives a fair idea of the class of houses rented to mine employees on the Hazleton mountains.

Number of Houses.	Rental per Month.	Number of Houses.	Rental per Month.
4	$1.00	20	$3.75
29	1.50	348	4.00
44	2.00	24	4.50
25	2.75	28	4.60
10	3.00	45	4.75
13	3.25	131	5.00
10	3.50	119	5.50
Total houses rented.		850	

According to the assessed valuation of these houses 42 per cent. of them was rated at from $20 to $50 ; 14 per cent. at $100 ; 8 per cent. at from $150 to $175 ; 3 per cent. at $200 ; 16 per cent. at $300, and the remaining 15 per cent. at $400 and over. In the last group are found the houses in which the foremen, clerks, officials, etc., of the company live. In the other groups are the houses in which the workingmen live and the best are assessed at $300. The rule in these communities is to assess real estate at one fourth its market value, so that the market value of the best of these houses would be $1,200 for a double house, while half of them fall below the $400 mark. The company realizes 14 per cent. annually on the best houses, but its returns from the miserable old shacks which were built from 40 to 60 years ago is much higher. A computation of the annual rentals and the market value of all the houses makes the returns of the company between 18 and 20 per cent. per annum.

The following is a classification of houses owned and controlled by the Philadelphia and Reading Coal and Iron Company:

MONTHLY RENT.

Number of Houses.	Number of Rooms in House.	Highest.	Lowest.	Total.	Average.
6	2	$3.00	$1.50	$12.50	$2.08
469	3	4.00	1.00	1,319.40	2.81
1,115	4	6.00	1.00	4,218.23	3.78
269	5	9.00	1.00	1,231.30	4.58
85	6	9.00	1.25	430.75	5.07
89	Over 6 and up to 12.	16.00	4.00	722.00	8.11

In November, when this statement was compiled, the total number of houses owned and controlled by the company was 2,217; number occupied by employees, 1,793; number occupied by others, 240; number occupied by indigent persons and widows for whom no rent was charged, 47, and 137 of the houses were vacant. The total monthly rentals of the company amounted to $7,934.68. Of the employees of the Philadelphia and Reading 7.3 per cent. lived in company houses, while 3,284 or 13.3 per cent. owned their homes. Of the above houses about 50 per cent. of them rent from $1.75 to $3 a month; 40 per cent. from $4 to $6, and the remaining 10 per cent. for $7 and over. When one of the officials of this company was asked why the company did not plaster the houses, he replied: "It's no use, it wouldn't last a year." Small repairs to windows, doors, etc., are left to the tenants themselves in houses which do not bring more than $4 a month rent. Carpenters are sent around the houses to do needed repairs about once every three years. Seventeen per cent. of the total rentals is spent in repairs. The Philadelphia and Reading seldom ejects a tenant because of arrears in rent. Some tenants are known to have owed the company over four years' rent and still they were permitted to occupy the houses. Rents are generally higher under the individual companies than under the large companies. The class of houses for which the Philadelphia and Reading Company charges $5 and $6 a month is $7 and $8 under individual companies. Rents of houses have been reduced as the properties have passed into the hands of large corporations. The Lehigh and Wilkesbarre follows the rule of ejecting those who will not pay the rent for two consecutive months. This company rents about 450 houses on the Hazleton mountains, charging $2.50 for shanties and $4 and $5 for houses of 4 and 5 rooms. The Van Wickle estate rents about 300 houses and charges from $3 to $5 a month, according to grade. In Milnesville, this company gave its tenants free coal, which was not done to the employees of the same company living in its houses at Coalraine and Evans. Carpenters are sent by this company around the property once a year to do any repairs which might be needed. Its rule also is not to collect rent from widows of men killed in its mines.

All through these coal fields there is no fixed rule for rentals. Under the same company, houses vary in rent when it is hard to see why the difference is made. On the Hazleton mountain the rule generally is that $1 a month is charged per room. Great variation prevails. In Silver Brook 3-room houses rent for $4.50, and houses with 5 rooms and an attic for $7 a month. The Lehigh Valley charged $6 for 4-room houses, but it also sold 12 tons of coal per year to its tenants for $1 a ton.* Rents under the Philadelphia and Reading are regulated according to floor space and convenience to market. They range from $1.75 to $15 a month. Under the Millcreek Coal Company commodious houses are furnished their employees for from $6 to $8 a month, while another individual company not far distant charges from $5 to $6 for shells which do not keep out the wind and drifting snow of winter, and around the base of which the tenants pile ashes to keep out the wind which comes under the floor between the frame and the blocks upon which the houses rest. No cellar, no foundation, no plastering, no paper, no ceiling, simply the frame with rough hemlock

* The company now charges its employees market rates for all the coal they buy.

One of the Best Company Houses in the Coal Fields. (A double dwelling.)

boards nailed upright and strips fastened over the joints. On a cold winter's night, one of the fathers dwelling in one of these houses, was awakened by the cry of his children who were cold and could not sleep. Two stoves were kept burning, but the cold could not be kept out by such a dwelling on a blustering winter's night on the mountains of Schuylkill.

The lower grade of company houses, as above stated, are today occupied by Sclavs and in many of them there is crowding. Under the Dodson Coal Company in Morea there are two groups of dwellings. The one on the northwest of the colliery comprises 106 dwellings which are inhabited by English-speaking people, and have on an average 5.1 to the house. On the southeast side is a patch of 30 dwellings wholly tenanted by Sclavs and having between 9 and 10 persons to the dwelling. Here, when the colliery is in full operation, it is nothing unusual to have from 6 to 8 boarders in the same house. Every morning more adult mine employees respond to the gong of the breaker from the 30 dwellings on the east side than from the 106 on the north. The same is true in other mining camps. The Lytle Coal Company rents 140 houses; many of these have been built recently and are good specimens of workingmen's dwellings of 5 and 6 rooms, garret and cellar, which rent for $7 and $8 a month. But the older dwellings occupied by Sclavs rent for from $2 to $4 and are for the most part fit only for the fire. These people pay high rent if computed by the floor room they possess. In Lattimer there are two mine patches. In the one to the east of the mines substantial houses of 4 rooms, a garret and a cellar are rented for $5 a month, but on the west of the colliery are old shacks, comprised of only 2 rooms (10 ft. by 12 ft. and 10 ft. by 7 ft.) for which the company charges $2 and $3 a month. One of these tenants, an Italian, desiring more room, built an addition at an expense of $40 for which the company allowed him nothing.

Many companies in recent years have built houses for their employees but in every instance they are far better dwellings than those which were put up in the early years of mining.

This is true of the houses built by the Coxe Bros. at Drifton, by the Markle Co. at Japan and Ebervale, and by the Lytle Coal Co. at Minersville, while some of the cottages put up by the Philadelphia and Reading Coal and Iron Company, are model dwellings. "Bosses' Row" in Maple Hill, erected by the latter company, contains commodious dwellings, but the agent having these in charge said that the investment did not net the company 6 per cent. per annum. One thing is patent, the better houses erected in recent years by these companies do not pay so large a dividend as the miserable shanties built half a century ago.

The following is a list of houses erected by the above company in recent years for some of its employees.

Place.	Number of Houses.	Average Cost of Building.	Rooms in House.	Rent per Month.
Ellengowen.	5	$984.48	8	$9.00
Eagle Hill.	10	811.33	6	6.00
Silver Creek............	13	1,355.87	8	8.50
Maple Hill............	16	1,158.04	9	9.00
Shenandoah	10	838.00	7	9.00

But in building new houses we do not find uniformity. An individual company in Schuylkill county turned an old barn into three dwellings for the Sclavs. The men called it the "barracks." Each of the houses had three rooms, two of which were about 16 feet square; from this converted barn the company reaped $18 a month rent.

The shells and shanties in which so many of our people live are cold in winter and warm in summer. When some of these people were asked how they managed to sleep in them in the summer months they replied that they did not sleep in them. Some sleep on the roof. Most of these people build some kind of a shed in the open lot and sleep there. Mosquitoes trouble a little, but these are more tolerable than the enemies within. A mason who had occasion to repair a chimney in one of the houses occupied by a Hungarian said, "You could shovel them." There is no denying the fact that even the rigor of winter cannot destroy all life in these shells where dirt and filth accumulate.

We have spoken of the company houses in the Middle and Southern coal fields especially, for in these sections of our territory this evil is most prevalent. There are miserable shanties in the Northern coal fields, but their number is insignificant as compared with that of the above-mentioned regions. Scranton Flats near Wilkesbarre, and rows of company houses in Plymouth and Nanticoke are dreary and wretched. The Delaware, Lackawanna and Western rents 284 houses, the monthly rent ranging from $2 to $8. The Hillside Coal and Iron Company rents 107 houses, the average rental per month being $5.40 and the average number of rooms to the house 5.8. The Delaware and Hudson rents about 150 houses which have on an average five rooms and the rents range from $4 to $8 a month. Many of the individual companies in the Wyoming and Lackawanna valleys rent houses but, as above stated, the evil does not prevail as it does in the Hazleton and Schuylkill regions. Greater opportunity has been given to individual enterprise, and a far greater percentage of the mine employees in the Northern coal fields own their homes than in the Middle and Southern.

A system of leasing ground for building purposes also prevails under many companies. The Lytle company has 178 such leases, the lessor paying from $1 to $10 a year for the use of the land according to the size of the claim. Some of the companies on the Hazleton mountain do this. Calvin Pardee does so, charging fifty cents a month rent for the land which is ample for a dwelling and a garden large enough to supply the family with vegetables. The same is done by some of the companies in the Northern coal fields. The Erie Co., the Delaware and Hudson, the Susquehanna Coal Co., etc., lease land on which their employees build houses. The Philadelphia and Reading, however, does this on a larger scale than any other company. In the Tremont valley, many of their old employees have held claims for years and their small farms are a source of profit and pleasure to the mine workers. The same is true of other sections where small farms are held by men who earn their chief subsistence by cutting coal. In the neighborhood of Silverton, where the same practice prevails, it is nothing unusual to find families who have resided in the same place for the last fifty years. During the strike of 1902 many of these employees of the Philadelphia and Reading were better able to carry on the struggle than their brethren of populous towns and cities, for the reason that they had the produce of their small farms to fall back upon.

We have spoken of the miserable shacks found in many mining camps. It must not be supposed that these patches are unhealthy habitations for men. It is difficult for the tenants to make these places unsanitary. They are generally perched on the mountain side; the double houses are generally detached and form long rows, while on all sides lies the open mountain. There is no restriction for room, while an abundant supply of sunshine and pure air is found. No healthier spots for human habitations can be imagined than these mining camps among the mountains, and whatever diseases prevail are largely due either to the lack of ordinary precaution or the total neglect of the rules of sanitation. Gross negligence is often observed in this regard. On one of our peregrinations in the summer to a camp occupied wholly by Sclavs, one of the directors of the company advised us to eat our supper before we visited it. The advice was judicious for the rancid stench that surrounded those dwellings clung to our nostrils long after we left the camp. Every tenant did as the dwellers in the kitchen middens were wont to do: the offal, the dish-water, the suds from washing, and even the excretions of the human body were thrown from the door or window of the dwelling, and all this filth in the glare of the summer's sun, gave rise to an effluvia that was sickening. What wonder is there that under such conditions the death-rate of the slums is found here on the mountain heights?

Another 84 per cent. of the mine employees live in houses owned by private parties. Many of these are owned by the miners themselves. The houses have on an average six rooms. They are built with better taste than the uniformity which dominates company houses, while they are invariably painted in a light cheerful color, which stands in striking contrast with the dull red paint which generally adorns the company house. What a contrast there is between the niggardliness of coal companies when they erect homes for the people, and the taste of the men themselves when they put up their own dwellings! The dull monotony of a mining village is oppressive. Sixty double houses may often be seen uniformly built and placed in two or three rows; all of them with slanting roof over the kitchen; no porch, not the faintest attempt at decoration in any part of them; each block speaks of parsimony in its construction; the impression comes with irresistible force that these houses were built for rent. The homes owned by mine employees differ; they have a porch, and the windows and doors have something to break the monotony, while invariably there is a side door and small porch. Then the house has as many rooms on the second floor as on the first, which give the dwelling a more symmetrical appearance. The gable is generally adorned, while there is invariably a coping crowning the roof. Two or three harmonious colors are generally used in painting which give the house an appearance of comfort and liberality. These houses rent at from $6 to $10 a month, according as they are located in greater or less proximity to populous towns or cities.

As before stated, better opportunity for individual ownership has been afforded in some sections of the coal fields than others. The Pennsylvania Railroad Company has not built many houses in Williamstown and Lykens, in Dauphin county. It only has about fifty houses all told in both places, which are built for the convenience of employees who are thus in close proximity to the collieries and can be quickly summoned in case of emergency. This company also, in former years, refused to join the other railroads, commanding the tonnage of the coal fields, to curtail production, so that for the last 20 or 25 years its employees have had steady work and have been able to earn good wages. The result is that the towns of Lykens and Williamstown have been mostly built by mine employees, and a larger percentage of these men own their homes than in any other section of the coal fields.

In the Panther Creek Valley, where the Lehigh Coal and Navigation Company carries on its operation, while the company rents 330 houses to its employees, ample opportunity has been afforded to the enterprising and thrifty worker to secure a lot and build a home of his own.* The result is that out of 3,043 families from whom the company gets its employees, 870 or 27 per cent. of them own their homes. In sections of Schuylkill and on the Hazleton mountains individual ownership is out of the question. The coal companies own all and hold all. In Cass township, out of a total of 1,170 taxable persons, no miner or laborer owns any real estate. It is a significant fact that this township was the chief center of activity of the Mollie Maguires, and in recent years it has furnished more paupers, according to its population, than any other section of Schuylkill county. In the township of Mahanoy not one of the 1,200 electors there owns real estate. In Hazle township there are over 2,500 taxables, nearly all of which are in the hands of the companies. In the towns of Freeland and Hazleton individual enterprise has been given free scope and two thriving towns are the result.

This restriction of individual efforts in securing homes is apparent in the number of hired houses in the three counties dependent on the anthracite coal industry. In Lackawanna there are 56 per cent. hired houses; in Luzerne county 64 per cent., and in Schuylkill 63.6 per cent., which shows that 8 per cent. more families live in hired houses in the two latter counties than in the former. Taking Lackawanna county we find that it compares favorably with other counties as to the number who own their own homes; in Lackawanna 44 per cent.; in Berks 41 per cent.; in Susquehanna 47 per cent.; in Clearfield 49 per cent., and in Westmoreland 41 per cent. In the computation we have left out the farms and made the comparison only of "other houses" as given in the census. In all Pennsylvania 41.2 per cent. of the families owned their homes,

* In Lansford, land is at a premium to-day. The few lots still available for building purposes cost from $30 to $50 foot front, which accounts for the fact that few Sclavs have built homes in this borough.

of which 26.8 per cent. were free and 14.4 per cent. encumbered. In Northumberland county, of 11,293 adult mine employees, 1,488 or 13 per cent. owned their homes. Of the 1,488 homes, 854 or 57.4 per cent. were mortgaged and 634 or 42.6 per cent. free. This gives about 31.4 per cent. of the families of mine workers in this county who own their homes, of which 18 per cent. were mortgaged and 13.4 per cent. free.

The Delaware, Lackawanna and Western found that 28 per cent. of its employees owned real estate. In a computation made by the Delaware and Hudson Canal Company the percentage was about 29, and their holdings aggregated $1,322,161. The same spirit of enterprise is seen everywhere in the Wyoming and Lackawanna valleys. The mine workers are given an opportunity to buy a lot and build a home, and they do so to as great an extent as any class of employees of equal earning capacity. The Sclav has shown remarkable avidity in this regard. In the mining towns north of Scranton, including Dickson City, Priceburg, Throop, Blakely, Olyphant, Peckville, Ridge, Jessup and Archbald, there are 4,000 houses owned by individual property owners; of these 1,700 or possibly 43 per cent. are owned by Sclavs or the non-English-speaking population, all of which has been attained in the last ten or fifteen years. This readiness of the Sclav to secure a home is apparent all through the Northern coal fields and it is a loss to society that greater opportunity is not given these men in the Middle and Southern coal fields to secure homes and attain the blessings which come by individual ownership of part of the soil. The Philadelphia and Reading has done much by its employees, but its refusal to sell land to them has materially reduced the percentage of its workmen owning real estate, while the policy of the individual owners on the Hazleton mountain has been to exclude all private ownership from its holdings. One of the reasons the Philadelphia and Reading advances for not selling building lots is, that the surface when undermined is liable to cave in and it cannot afford to assume the responsibility of keeping the surface intact or pay damages to property owners in case of a cave-in. There is reason in this argument. The board of directors of Mahanoy township in 1900 sought a site for its high school building, and the principal reports: "Ramified by a net-work of mines it is simply impossible to secure a safe and desirable location for a school building; . . . over two miles of ground we sought a suitable location and were at last brought to the alternative of abandoning the proposed building or accepting the present undesirable site." Still the fact remains that the company houses have stood for over half a century, and some system should be devised to give the mine workers of these localities an interest in the soil.

The Condition of Woman Therein.

In every institution there must be an ultimate source of authority, and in every family the question "who is chief" must be settled if the home is to be one of peace and order.

In the miner's home there is little room for sentiment, and among these practical people the husband is generally lord of the home, and the wife must hold herself in subjection to him. Division of labor is practiced to the strictest detail. The husband is the bread winner. All work in the home belongs to the wife, and if the husband lights an extinguished fire, or cares for the stoves, or "minds" the baby or aids in the preparation of the meal, he "helps" his wife and speaks of work of this nature exactly as if he had given a helping hand to his neighbor. If the wife neglects the household duties, by not preparing the evening meal on time or not having the water ready for the daily ablutions, she is called to account and disciplined. The husband generally gives the wife his pay and expects her to make good use of it. Many of them hold their helpmeet to a strict account in the items of expenditure for the maintenance of the institution. As a rule the words of Napoleon are believed and practiced in the houses of the mine workers: "A husband ought to have absolute rule over the actions of his wife."

This has been the code by which the homes of the foreign born have been governed. Among the descendants of the foreign born it is not so religiously practiced, and the revolt comes from the female sex which demands the recognition of certain rights, more imaginary than real, and resents the lordship of the man of their choice. Under the old régime large families were raised and the parents attained a measure of domestic felicity seldom enjoyed under the new code, while virtually no divorces are known among the foreign born. The parties to the divorces granted in the courts of the mining regions are invariably native born. This possibly has much to do with the fact that the number of divorces granted in our State in the last five years has more than doubled, while the causes given for the suits are, for the greater part, trivial. A tendency lightly to regard the holy vows of matrimony grows to an alarming extent among the native born of foreign parentage. It bodes ill to the peace of society and strikes fatally at the family relation which is the basis of all that is good and honorable in social life.

The lot of the miner's wife is a hard one. The code of domestic ethics prevailing among these men largely savors of that of marriage by purchase. The wife is the property of her husband, to be used according to his will, and for which he paid a price. Among the mine workers large families are still the rule. Those who feel the effect of a rise in the standard of living, which carries an effect upon procreation, get out of the mines. The daughters of miners, whose tastes conform to the American standard of living, prefer to remain spinsters rather than join their lot with mine workers. This is the reason why in every mining town there is a class of females among the native born who lead a single life, while among the foreign born engaged in the mines an excess of bachelors is found. No one familiar with the drudgery and toil of the miner's wife can say anything to these women who prefer single life.

Take a day's round of toil in these women's lives. The miner gets up at 5 or 5:30 in the morning, and the wife must be there to prepare breakfast and fill the pail of the husband. The husband gone, she does a little house work before the children awake, who must be fed and dressed for school. When the children are gone, the nursling must be washed and fed.

The morning is spent and at noon the children again come from school and their wants must be supplied. In the afternoon the evening meal — the only cooked meal during the day — must be prepared and a plentiful supply of hot water for the daily ablutions of the mine workers. Between four and five the workers return home, and so do the children. Clean warm clothes must be got for the husband and the dirty garments laid aside for the morning. The tub is then removed and the hearth cleaned of suds before the family sits at table for the evening meal. Then comes dish washing. No sooner is this done than the children must be prepared for bed. The wife's work is not yet done. A patch is needed on the children's or husband's clothes and when all are asleep she plies the needle in order that those in her care may be fairly well clad. No wonder these wives and mothers are worn out at the close of day and long for rest. Add to this daily round of toil the washing and the baking, the shopping and the house-cleaning, not to speak anything of the night vigils they keep when some of their household are sick. There is but one pair of hands to do it all. Hired help is out of the question, for the wages of a mine worker will not permit it. Can we wonder that the homes and the children of these mine workers are not so clean as they should be? And is it strange that the daughter of ten years is kept at home to help mother before she has learned the rudiments of a common school education? Many of these brave hearts break down when their husbands are still in the vigor of manhood. They are worn out, their frames are shattered, they look prematurely old, and the causes are the burden of motherhood and the ceaseless toil of home. What is there of poetry and music in such a life, and native born girls of refined taste, who have seen it in all its dreadful reality, shun it as they would the galleys.

Much has been said of the hardship of a miner's life. The irksome and disagreeable conditions of employment and the hazardous and toilsome work of mine workers, have been set forth in the agitation for better conditions during the past years, and they deserve to be proclaimed. But in all the agitation nothing was said of the miner's wife, the hard conditions of her life and the burden she has to carry in the conflict of life. When a miner testified before the Anthracite Coal Strike Commission, and stated that his wife was then mother of ten children and that she was sent to the asylum, Judge Grey said: "No wonder she went to the asylum." No, the wonder is that so few of these burden bearers in the homes of the anthracite coal fields go to the asylum.

But the question comes, what can be done for these mothers whose burdens make them dead to rapture and despair? How are they to be relieved? There seems no door of relief as long as the husbands cling to the "wife-by-purchase" code of domestic ethics. The belief that the wife is the creature of the husband's pleasure is fatal to all attempts at ameliorating her position. Reason and common sense seem to suggest that the first step is to bring the men to feel and acknowledge that the woman has rights which must be observed and that to place upon her a heavier burden than she can carry must inevitably result in deterioration in both the body and the mind of posterity. The virtues of self-restraint and self-control are the great lessons to be learned by these husbands, and no efforts of unionism or legislature will permanently aid them unless these virtues are practiced. If the social status of the working classes is to be permanently improved, restriction of natality must have a larger place in their creed.

The women also deserve attention. The "absolute sovereignty" of the husband may contribute to domestic peace, but in the home the woman is queen and the permanency of her power depends upon the efficiency with which she is able to discharge her duties.

The great need of our mothers is proper training in the domestic arts. That will be a greater adornment for her and a greater blessing to society than any accomplishment which demands a more prominent sphere of display. To teach her plain cooking, the nutritive qualities of common articles of diet and the best way to prepare them for the nourishing of the body; to teach her plain sewing so that with neatness and economy she can care for the clothing of her household; to teach her how to care for children, so that a decreased fecundity will not mean a decrease in population; to teach her that there is no poison as virulent as that which man's own filth generates, and that dirt and death go hand in hand; to teach her that cleanliness, fresh air and sunshine are prime conditions of life and happiness — these are the things that need to be taught to the wives and mothers in the anthracite coal fields of Pennsylvania, and perchance to the wives and mothers of the working classes in general. The chairman of the committee on Preventable Diseases of the State Board of Health said in 1897, that 6,000 children die annually in the State of Pennsylvania of preventable diseases, and that more than one third of all the children born in this Commonwealth die under five years of age. We have seen that the death-rate among infants in these mountains equals that of the slums of crowded cities, while it is exceeded only by that of the colored population of our country. Much of this waste of life is due to ignorance and shiftlessness and dirt. If mothers' meetings are needed anywhere, it is here. Nowhere can the teacher of practical hygiene and the elements of sanitary science be of greater service; while the moralist who teaches temperance, self-control and right relations has a rich field for his practice. Among most of these people the desire to rise in the social hierarchy is apparent. They are susceptible to better things, and especially so are the women. As a first condition of improvement they should be made to feel the close connection between dirt and degradation. It is far closer than most people are willing to believe and recognize. Whoever is subjected to a condition of life where cleanliness is not practiced will find it almost impossible steadily and surely to improve in his moral tone.

Another thing, better homes should be placed at the disposal of families who live in miserable shanties. As long as mine workers live in these wretched hovels, it is hopeless to expect a better type of womanhood and a more exalted idea of the relations of family life.

The Rights of Children.

In the family each of its members has rights which public opinion, custom and legislature prescribe. The husband is lord of the home among these workers and yet he cannot maintain discipline as men were wont to, namely, by force. The knout is not used by any here, while the rod is relegated to oblivion. There is a chivalric spirit among these men, that to strike a woman is cowardly. Some men do beat their wives, but the force of public opinion is such that it never fails to find some way of expressing itself, and if the offender does not amend his ways, intrepid amazons take the case in hand and discipline the culprit. Between husband and wife in mining communities there is observance of mutual rights, and generally speaking the woman recognizes the authority of her husband and insists not on the acknowledgment of equal rights, which invariably leads to conflict.

There are many children in the families of mine workers whose rights, as far as they are still left in the hands of the parents, are observed. Parental love is strong in our people and marriage without children is regarded as unfortunate. It is only when the descendants of foreign born parents fall under the influence of a rising standard of living and their felt want far exceeds their real want that fecundity is decreased. It is curious, however, that the average family in our communities has a greater welcome for boys than girls. This is carried to a greater extreme among the Sclavs than the English-speaking. When a boy is born in the home, the christening is an occasion of great rejoicing and the "kum" assembles and spends the evening in feasting and dancing; but if it is a girl the occasion is passed by in silence. This same sentiment prevails generally among the English-speaking. One of these, when told that the child born to him was a girl, refused to speak to his wife for several days. Another beat his wife because she gave birth to the seventh girl. Preference for male children has characterized barbaric nations. The preference of our people is due to economic causes, for many of the parents look upon the growing boy as a promise of future aid in the conflict of life.

In these homes parents are indulgent to their children. Most of them feel their moral obligation in rearing them and practice moral restraint in their presence. There are some wife deserters. They are, as a rule, lazy fellows who will not support the brood they have brought into the world, but, departing, they leave them to become a public charge. Many such children are in the orphanages of our counties. But, generally speaking, these mine workers work hard for the maintenance of their offspring, and are anxious to clothe and feed them well. They are not always able to do so because of the frequent additions to their families. So short a period as 15 or 18 months between births is detrimental to the physical well being of children. They cannot get the care and attention they require, and although parental love is strong yet there is a limit to human strength and patience. Children will be physically and intellectually better, when the period between births is extended to three or four years. Frequent births must result in enfeebled bodies, weak intellects, and dispositions warped by the frequent outbursts of passion in an overworked and underfed mother. Under the present nemesis of ignorance and passion, the rights of babes cannot be maintained, no matter how strong may be the love of parents for their offspring. Here again wrongs are committed in ignorance, and only the dissemination of knowledge can effect the cure.

The homes of many of these mine workers are far from being suitable places to impart the first impressions to the receptive minds of the children. Not only are the ordinary parents ignorant of the laws of mental development, but they are also depraved and vicious, and the effect upon the awakening mind of the child is often tragic. Home is the best place for a child, no matter how humble it may be, but to keep a child for the six first years of its life under the sole care of parents wholly incompetent to direct and mould the awakening mind, is mischievous and cruel. The State, which has shown anxiety for the proper training of children, should extend its power and bring the child, which spends only an average of five years in school, under its control at an earlier age. It means the extending of the kindergarten system to all our communities, which is devoutly to be desired. Letourneau said: "Our actual family circle is most often very imperfect; so few families can give, or know how to give, a healthy physical, moral and intellectual education to the child, that in this domain large encroachments of the State, whether small or great, are probable, even desirable. There is, in fact, a great social interest before which the pretended rights of families must be effaced. In order to prosper and live, it is necessary that the ethnic or social unit should incessantly produce a sufficient number of individuals well endowed in body, heart and mind. Before this primordial need all prejudices must yield, all egoistic interests must bend." It must have been the incompetence of the ordinary mother properly to judge of the best interests of society that led the divine Plato to believe that the mother, after the birth of the child, should be under the direction of the State. No one to-day believes in any such encroachment upon the rights of motherhood, but it would seem just that the child should be placed under competent teachers before the age of six.

The miner's home also is rarely large enough to afford that protection to youth which the laws of decency and morality demand. We speak elsewhere of ugly furniture and unsightly pictures; of nasty papers and trashy books; of the absence of means of culture and refinement. What we refer to here especially is the daily custom of ablutions, which are practiced on the hearth in the presence of children of both sexes. Every one employed in the mines must take his bath at the close of each day's work. The homes of mine employees know nothing of the luxury of a bath-room. A tub is used, which is invariably placed in the kitchen. During the task of washing, the children pass to and fro, and are spectators of the nakedness of the bather. This practice is fatal to modesty between the sexes and the sense of decency. In the ordinary miner's home there cannot be seclusion, and so accustomed are they to this practice of nudity that little attempt is made at seclusion. In the summer months many homes have outside shanties where the daily ablution may be performed with a degree of privacy. Then, however, the members of the household are out in the sunshine and pure air. But in the winter season the children cannot go out into the cold; they stay indoors and are daily witnesses to the father's or the brother's bathing. Those who live in a six-room house can, if anxious to do so, escape the deterioration incident to such a practice, providing they are willing the children should use the front room, which is the best furnished in the house and generally called the parlor. But even if a wise and sensitive mother is anxious to protect her children against the evils incident to familiarity with nudity, they are liable to meet the same in their neighbors' houses. It is difficult to see how this blunting of the sense of decency in the female sex can be avoided. To erect bathrooms in the homes of mine employees is out of the question. A law was passed in 1891 requiring the companies to erect wash-houses at each colliery where the employees could perform their daily ablutions. This was chiefly designed for men working in wet places, who were obliged to walk home in their wet garments in winter. Few of the men avail themselves of the provisions made. They have their reasons. Wet garments must be dried for the following day, and the company has no one present whose business it is to see that the miners' clothes are dried. If he leaves them near the stove or steam pipes, they will be stolen before the morning. The mine employees will continue to wash in their homes, and as at present done, the system is demoralizing. Daily exposure in tubs on the hearth where children receive their first impressions of propriety and decency and where young girls come to womanhood, must affect their morals. Many parents could do much to protect their children, if houses large enough were placed at their disposal, and they were fully conscious of the evil incident to the present custom. But as long as many of our people live in houses of only two rooms, it is impossible for them to practice seclusion, however much the prudent mother may wish it.

Another evil of houses with inadequate accommodations for ordinary families is the lack of bed-rooms which are separated the one from the other. How can modesty and decency be practiced in a house with one bed-room? We have seen homes where the only bed-room had in it four beds, in which the parents and the children slept. Under such conditions can we expect children to grow up virtuous, modest and pure? If abominations exist such as are practiced among the heathen, it ought not to be strange. Sanitary and moral conditions can never be satisfied under such a system, and the rights of young children to wholesome examples in decency and self-respect cannot be maintained. It is a cruel wrong to growing manhood and womanhood to subject them to such environment and expose them to such temptation, and if incest, illegitimacy, juvenile prostitution, idleness and disease prevail, society pays a dreadful price for tolerating such conditions.

Every child born into the world ought to have a plentiful supply of nourishing food, warm clothing, adequate shelter, wholesome example and proper tuition. Under conditions which prevail in many mining camps these rights cannot be secured them, and those who have the interest of society at heart ought to see that the rights of children are more sacredly and religiously maintained.

The Need of Better Homes.

In the towns and villages of the anthracite coal fields there is not much crowding. The 400,000 people are, for the greater part, scattered in small settlements of from a few hundred to 20,000 population. In the 1,700 square miles of territory occupied by the coal fields there are only two cities of any size, namely, Scranton and Wilkesbarre, and even in these there is little congestion. The 102,000 inhabitants of Scranton are widely scattered over an area of about 16 square miles, so that the residential portions of the city have the appearance of suburbs as compared with congested districts in large cities. Wilkesbarre with its 51,000 is the same. Outside these two cities, the 127 municipalities which are located in the anthracite coal fields are classified as follows as regards population:

Of less than 1,000 population	26
Between 1,000 and 2,000	22
Between 2,000 and 3,000	29
Between 3,000 and 4,000	9
Between 4,000 and 5,000	15
Between 5,000 and 6,000	8
Between 6,000 and 7,000	3
Between 7,000 and 8,000	1
Between 9,000 and 10,000	1
Between 12,000 and 13,000	4
Between 13,000 and 14,000	4
Between 14,000 and 15,000	2
Between 15,000 and 16,000	1
Between 18,000 and 19,000	1
Between 20,000 and 21,000	1
	127

Of these townships and boroughs, none can be said to be congested save Shenandoah and Mahanoy City, and of these two the former is the worse. The town of Shenandoah is located on a knoll which would, with ordinary precaution, afford admirable opportunity for sanitary homes. The town, however, occupies about one fourth the area it should, and is surrounded by coal properties which are either in litigation or in the hands of parties who will not sell the surface for building lots, so that all available ground has long been appropriated and real estate has been for many years at a premium. Here building lots of 25 ft. by 75 ft. have four dwellings upon them, and between these are the vaults for the use of the tenants. In summer time the stench from these is sickening, and often have we heard the tenants complaining that they cannot open the windows facing these pest holes either night or day, while members of the family have in the morning spells of nausea which destroy appetite, and frequently bring on sickness. The towns of Shenandoah and Mahanoy City are also built on the coal basin. Many houses are built on the rocks which pitch at an angle of 40 or 50 degrees. On the higher elevation the closets are built, and at a distance of 30 feet the cellar of the dwelling is dug. The floor of the cellar is frequently lower than the bottom of the vault, and the seams being slackened by mining operations, the contents of the vaults sometimes enter the cellar—a nuisance which brings death to those tenants most susceptible to disease. Infant mortality in these two towns is very high. When we consider the pestiferous vaults crowding the few feet of yard where these people dwell, the wonder is that the mothers carry not fatal bacilli enough to kill every nursling in the home. The only redeeming feature in these towns is the open mountains which surround them on all sides, where the children can run in summer days and breathe the pure air and bathe in the warm sunshine. Both towns are also supplied with good water, although the supply furnished by the borough of Shenandoah is not adequate to meet the demand, which is a source of great inconvenience in a mining town where a plentiful supply of water is so essential to the comfort and convenience of the workers.

In both these towns the local boards of health ought to be more vigilant and strict, and the greed of landlords ought not to be permitted to stand in the way of the health of the people. The words of Dr. G. G. Groff: "The waste from any animal's own body is the poison which it should most dread" should be remembered. The greatest menace to our towns arises from lack of adequate precaution in the matter of vaults and sewers, and the indifference or total absence of health officers in most of them accounts for the presence of many nuisances which need abatement. There is no justification for the high death-rate among infants in these mountain towns, and the rights of children born into the world will never be adequately protected until public sentiment is so awakened that the many nuisances now existing are removed. It can only be done by putting social well being above personal gain, and placing the affairs of our boroughs in the hands of men who regard their trust as paramount to any personal interest.

We have spoken specially of Shenandoah and Mahanoy City, but the evils complained of are generally in towns of from 3,000 to 20,000 inhabitants. A visit to Lansford, Mt. Carmel, Olyphant, Plymouth, Edwardsville, Nanticoke, etc., will reveal the same evils. Many of the people in these towns are not conscious of the deadly poison arising from unclean habits, and while this ignorance and filth prevail the angel of death ever hovers around the lives of innocent children.

We shall speak in a later chapter of the intemperance of mining towns, but in this connection a few words will not be amiss. The connection between the poverty of the home and the saloon has long attracted the attention of students of society. Miserable homes drive men to saloons. As long as men are compelled to live in houses where there is no room save to sleep and eat, the saloon will have an irresistible attraction for them. Increase of wages and shorter hours of labor will not benefit men who live in shanties and dwellings owned by companies. Give them more wages and more leisure, what incentive is there for them wisely to use their extra time and money? In most instances it will only mean more drink and more time spent in the saloon. The first step in the necessary reform in these communities is better homes for at least 20 per cent. of our people. A healthy and commodious home, such as working men deserve, will mean fewer saloons, greater health, and a larger measure of happiness. Then opportunity should be given the men to buy the homes at a fair price and on reasonable terms. Society as well as the companies will gain thereby. Personal possession brings with it a sense of responsibility and social worth, while it materially enhances the productive capacity of the individual. Place some object before the ordinary worker worthy of his best endeavor, and you increase his self-respect and independence, and he becomes a better man. Deprive him of all opportunities of self-improvement and self-advancement, and the possibilities of raising his social status will be forfeited. Society is stronger according as its members form a gradually ascending scale in personal possession of material good. No society can ever be firmly fixed where extremes of poverty and riches meet. A wholesome distribution of riches amid the various grades of society means health and vigor in the social group; amassing riches in one family by eliminating all opportunities of personal acquisition of property by employees must result in social disintegration and personal deterioration. If these few principles were acted upon by the coal operators of the anthracite industry, we believe greater peace and good-will would exist between employers and employees, and we are confident that the saloon evil would be greatly eliminated. If the coal companies that rent miserable houses to their employees, were to abolish these as speedily as they could and build homes for the people which would secure domestic privacy — the foundation of morality; sanitary conditions — the mainspring of health; and comfort, convenience and attractiveness; and offer them for sale on favorable conditions, they would confer a far greater blessing upon their employees than any advance in wages or improved conditions ever can. Company houses that are commodious — and there are many of these — could be sold outright to those willing to buy, and we are sure that ultimately the companies would gain far more in the improved personnel of their employees than they lose in rentals.

It is of importance to society that the anthracite mining operations should pay a reasonable dividend to capitalists and a fair wage for labor. But there are considerations far more important than the economic in the group we study. We have here a mass of raw material which forms one fourth of our population. Every one of the immigrants has come to our country in the hope of improving his material condition. By no lever can these men be raised in their social status so rapidly as by this anxiety for acquisition of wealth. Whenever the opportunity is afforded them, they have readily taken hold of land and put up a house. Should not considerations of patriotism lead our captains of industry to give all opportunity to these immigrants to put their extra earnings in houses, where their acquisitiveness may be satisfied and their pride in personal possessions be gratified? Dividends are important, but the development of manhood and womanhood is of far greater importance. It has long been proved, and especially so from the experiments of Robert Owen, that no investment pays capitalists so well as that made for the comfort and elevation of their employees; and no curse is so sure and strong as that which falls upon capitalistic greed, which concerns itself only with returns on its capital regardless of the physical, intellectual and moral conditions of the employees whose labor fructifies the capital they possess.

CHAPTER VI.

OUR EDUCATIONAL APPARATUS.

1. Schoolhouses in Mining Towns. 2. The Men and Women Who Teach. 3. The Boys and Girls in School. 4. The Boys in the Breakers. 5. Can Our Educational System be Improved?

Schoolhouses in Mining Towns.

There are in the anthracite coal communities 2,429 schoools where the children are trained for from seven to ten months in the year. Each month must contain not less than twenty days' schooling so that the pupils are taught each year from 140 to 200 days. Taking 107 municipalities in these regions, we find the following classification as to the number of months taught each year:

2	Municipalities	for	7	months.
8	"	"	8	"
81	"	"	9	"
16	"	"	10	"

This gives an average of nine months or 180 days' schooling to the youth of these communities. The average number of months' teaching in the counties in which lie the coal fields is as follows:

			1901.		1902.	
Schuylkill	County	(Anthracite)	8.74	months.	8.72	months.
Luzerne	"	"	8.67	"	8.56	"
Lackawanna	"	"	8.35	"	9.44	"
Columbia	"	"	7.65	"	7.70	"
Carbon	"	"	8.30	"	8.25	"
Dauphin	"	"	8.47	"	8.45	"
Northumberland	"	"	8.28	"	8.62	"
Susquehanna	"	(Agricultural)	7.39	"	7.35	"
Clearfield	"	(Bituminous)	7.23	"	7.32	"
Westmoreland	"	"	7.50	"	7.35	"

For all the counties of the State, inclusive of Philadelphia, the average for 1901 was 8.28 months and for 1902, 8.32.

From this it will be seen that, as far as length of school term is concerned, the children of the anthracite coal regions enjoy better privileges than those of agricultural communities and of the bituminous coal fields, while they nearly equal the advantages, in length of school year, which the sons and daughters of residents of large cities enjoy.

The total valuation of the school properties in the territory under consideration is $5,858,160.63, which is equal to 10.8 per cent. of the estimated value of the public school buildings of the State. The population of the anthracite coal communities forms 9.8 per cent. of the total population of the State, so that the percentage of school property in these regions is slightly in excess of the percentage of population as compared with that of the State.

The liabilities of the schools in these coal fields are $1,363,-587.10, which is about 12 per cent. of the total liabilities of the schools of the State, so that the proportion of school debts in our territory is in excess of the proportion of our population as compared with the population of the State. The indebtedness of our schools is not to be accounted for because of the tendency of school directors to place taxes levied for school purposes at a low figure. The law sets a limit of 13 mills, and out of 104 anthracite mining muncipalities 57 placed the tax at the maximum limit as prescribed by the State. The classifying of these muncipalities is as follows:

57 boards levied 13 mills.
12 " " 12 "
6 " " 11 "
8 " " 10 "
5 " " 9 "
5 " " 8 "
6 " " 7 "
5 " " 6 "

The boroughs of Pottsville, St. Clair and Tamaqua, together with the city of Wilkesbarre form conspicuous exceptions to the general rule of fixing the tax levy for school purposes. These levied for 1901, 4.5, 4, 5 and 4.5 mills respectively, and for 1902, 5, 5, 5 and 4.5 mills. Of course the rate at which prop-

The Ravages Wrought by Mining.

Scene of a Cave-in on Lackawanna Street, Olyphant, January, 1902.
(Four houses, a barber shop and a hotel were swallowed up within an hour of the first warning.)

erty is assessed in the various localities has a bearing upon the total amount appropriated by local taxation for the maintenance of the schools, and yet it is significant that over fifty per cent. of the boroughs and townships place the tax at the extreme limit as prescribed by law, while another twenty-five per cent. place it between ten and twelve mills on the dollar. In an agricultural area of nearly the same population as the anthracite coal fields, we find that the school liabilities are only 5.4 per cent. of the total liabilities of the State schools, while the percentage of population in this rural district is 9.9 per cent. of the population of the State. Taking an area in the bituminous coal fields of about equal population with that of the anthracite coal fields, we find its school liabilities only 5.6 per cent. of the total school liabilities of the State, while the population of the area is 9.4 per cent. of the whole population of Pennsylvania. Thus in both agricultural and bituminous territories the proportion of school liabilities is considerably lower than in our area.

During the year 1901 the total expenditure for the maintenance of our schools aggregated $2,242,548.61 or 9.7 per cent. of the total amount spent on the schools of the State. This percentage of expenditure is virtually the same as the percentage of our population. Of this sum the State appropriation amounted to $493,401.68 which is also 9.7 per cent. of the total amount appropriated by the Legislature for school purposes.

During the year 1901 there were 123,384 pupils enrolled in the schools of the anthracite communities, and the sum of $18.17 per capita spent in educating them during the year. The per capita expenditure for the whole of the State for that year was $19.64. The cost per pupil in Philadelphia was $30.88 and in Pittsburg $37.94. The expenditure per pupil in the anthracite mining towns and boroughs is larger than in agricultural communities. In Susquehanna county the cost per pupil was $13.37; in Berks county $16.26, while in the bituminous counties of Clearfield and Westmoreland we have an annual expenditure of $11.76 and $7.16 per pupil respectively.

The number of pupils per school in our territory is 50.8, and notwithstanding the fact that our towns and villages are scattered, they form communities so thickly populated as to give the majority of our children ample facilities for training in the common branches of education, while in nearly all our boroughs high school privileges are afforded them.

Thus the number of schools, the length of the school year, the total amount spent in erection of school buildings and in the maintenance of our educational system, the enormous debt incurred in equipment and in improvements, indicate that the pupils in our territory have greater educational privileges than those in agricultural and bituminous regions, while for the majority of our pupils there are educational advantages offered which are equal to any offered in second and third class cities in the country.

School buildings in our territory differ greatly in character. In the mining patches, frame buildings of the plainest possible type are the rule. Around them no attempt is made at beauty or ordinary adornment; no plot of green sod; no flowers, and the building generally needs paint and repairing. In the larger towns better buildings are found. Many of them are built of brick, but the majority are frame buildings. In some of these attention is given to beauty. Gravel walks divide plots of green sod amid which are found beds of flowers. The waste paper and useless stuff thrown by hundreds of pupils in the school yard are carefully removed. The fence is neatly painted, and the external appearance of the structure shows constant care and prudent expenditure of funds. Other towns which have large schools pay little attention to either the external or internal appearance of their schools. Externally, there is nothing to please the eye or suggest a beautiful thought or produce an agreeable impression on the mind of the pupil. On all sides one has the impression of carelessness and shiftlessness, and many parts of the structure show a wilful disregard to the best interests of the community from the sole standpoint of the preservation of property. Internally, most of the schools of mining towns and villages could be greatly improved. Generally no picture, no models, no flowers — nothing that makes the school-room attractive, is found. It may be partly due to the lack of funds, but the greater burden of responsibility for this barrenness in public school rooms rests with the teachers and superintendents of these schools. Little or no attention is given by many teachers — not to speak of school directors — to æsthetic environment in public schools. Were more attention given to this, both teacher and pupil would profit thereby; the teacher in having scholars more amenable to discipline; the pupils in receiving impressions of order, neatness and beauty which no person can impart by precept.

In the anthracite coal communities there is much that is ugly, repulsive, base, and depressing. The contamination of our streams, the black creeks full of water laden with coal-dust, the dismal acres where the refuse from washeries has long been turned — these make a dreary environment. Trunks of trees stand in valleys, veritable ghosts of stately pines which no more know spring-time and summer. In many places acres of culm heaps, which are the refuse of a century of mining, stand as black monsters defiling our fairest valleys; the huge black breakers and shafts enveloped in a cloud of smoke and steam and dust when in operation; the scores of mining patches where houses have been built with depressing uniformity, while around them are seen heaps of ashes, tin-cans, old bottles, empty beer kegs, etc. That is the environment of thousands of youths in the anthracite regions and it inflicts upon the man incalculable wrong which influences their whole life. Amid so much that is ugly and debasing, ought not the plastic minds of these children be brought in contact with one spot that is beautiful and serene, which would exert a holy influence upon their souls and stimulate their æsthetic sense? When the environment of the public school and the interior of the school room conform to artistic taste in the highest sense, then a sacred influence will work upon the awakening mind of the child, which will add dignity and interest to the specific work of the teacher. Its grace and suggestiveness will also do something to repair the wrong done the child by the neglect and cupidity of those responsible for the depressing environment generally found in mining communities.

Besides the public schools, there are many parochial schools in our large towns and cities, where hundreds of pupils are taught. The number of these schools, the valuation of their property, the number of scholars and teachers, etc., are not given in the annual report of the State Superintendent of Public Instruction. Schools of this nature are found in Scranton, Wilkesbarre, Pittston, Plymouth, Nanticoke, Hazleton, Mahanoy City, Shenandoah, Pottsville, Mt. Carmel, Shamokin, Olyphant, Carbondale, etc. Irish Catholic churches have many schools, but they are not so general as those planted by the Catholic churches maintained by the Sclavs. The Hungarians, Lithuanians, Slavonians and Polanders are sanguine upholders of the parochial school system, and they institute such a school wherever they are strong enough to put up a church edifice. The Ruthenians or Little Greeks form an exception to this. They invariably send their children to the public school, but every day insist upon their attendance in the church, where catechetical instruction is imparted to them. The total number of scholars in parochial schools in our territory, as found in the Catholic Directory, is 12,781, which is 10.3 per cent. of all the children in public schools in anthracite communities. In some of our towns from 25 to 30 per cent. of the children of school age are enrolled in parochial schools. The following table gives the number of these schools in the anthracite coal fields, together with the nationality of the scholars.

Nationality.	Number of Schools.	Number of Pupils.		Total.
		Boys.	Girls.	
Irish	22	3,961	5,156	9,117
German	7	528	540	1,068
Sclav	17	1,220	1,376	2,596
Totals	46	5,709	7,072	12,781

The number of teachers engaged in teaching is not given; in schools where the number is given there is one for every 40 pupils. If the same proportion holds good in all schools there are about 319 teachers employed, almost all of whom are sisters of various orders.

Many of the Sclav churches conduct their parochial schools in the basement of the church, while others have built a structure specially devoted to this purpose. Some of the Irish Catholic congregations have put up excellent buildings whose environment is admirably kept and a conscious influence is wrought upon the receptive minds of the children in attendance. The interior of the schools is also adorned by pictures of the saints which inspire reverence in the heart of the child.

Jewish congregations also have institutions where their young are taught. The children are regular attendants on the public school, but in addition to this they are daily required to attend the school at the synagogue, where they are trained in the faith of their fathers. Most of the Jews in the anthracite mining towns are of the orthodox faith, and strictly rear their children in the same. They, however, send their children to the public schools and only engage the private teacher to teach the thorah and give instruction in a few foreign languages.

In addition to these there are a few kindergarten schools in some of our towns and cities. In 1898 the school board of the city of Scranton appropriated $1,000 for kindergarten work. This movement soon grew and it now forms a part of the regular system of instruction. Nine such schools are now conducted in various sections of the city, giving employment to ten teachers. Outside the city of Scranton no municipality has incorporated the kindergarten work in its scheme of public instruction. The Presbyterian Mission among foreign nationalities has founded and maintains five kindergarten schools in the Wyoming and Lackawanna valleys, where children of all peoples and sects are welcomed and instruction is imparted them by song and drill, which are designed to develop in the child's mind patriotic sentiment and the sense of propriety.

In the Middle and Southern sections of the coal fields a few kindergarten schools were started by individual enterprise, but lack of patronage and of appreciation soon closed them. Free kindergartens are found in the towns of Hazleton and Pottsville. Such schools are sadly needed in all our towns and no more promising field is possible for individual or corporate philanthropy.

The Men and Women Who Teach.

The total number of teachers in the State in 1901 was 30,044; of these 9,194 or 30.6 per cent. were males and 20,850 or 69.4 per cent. were females. The average salary of male teachers was $44.14 per month, and that of female teachers $38.23. If we leave out the City of Philadelphia the proportion of male and female teachers is 33.8 per cent. and 66.2 per cent. respectively, and the average male teacher's salary $42.14 and that of the female $33.08. In 1902 there were in the State 30,640 teachers, 8,585 or 28 per cent. male, and 22,055 or 72 per cent. female; average salary for males $44.92 and for females $33.75. In Philadelphia the percentage of male teachers is 31 and of female 69, and the average monthly wage of male teachers $42.96 and of females $33.34.

The total number of teachers in the public schools of the anthracite coal fields in 1901 was 2,345, of which 493 or 21 per cent. were males and 1,852 or 79 per cent. were females. The average monthly wage paid to the male teacher was $68.16, and to the female teacher $40.59. The figures for 1902 in our territory vary but slightly from these. The only counties in the State paying a higher average per month to their male teachers are Allegheny and Delaware, while the average paid to female teachers is higher than that paid in 60 out of the 67 counties in the State.

The opposite table gives us the number of teachers in the six counties where anthracite mining is being carried on, the proportion of male and female teachers, and the average salaries paid them per month.

If we compare these figures with those of two agricultural counties, Berks and Susquehanna, we find that the percentage of male teachers in the latter is much higher than in the coal fields, and the average salaries paid both male and female

	County.	Total Number of Teachers.	Number of Male Teachers.	Number of Female Teachers.	Average Monthly Salary of Male Teachers.	Average Monthly Salary of Female Teachers.
Anthracite.	Schuylkill	595	176 or 29.4%	419 or 71.6%	$65.61	$40.96
	Luzerne	795	160 " 20.1	635 " 79.9	70.31	42.21
	Lackawanna	643	71 " 11.4	572 " 88.6	70.74	39.97
	Carbon	50	12 " 24.0	38 " 76.0	53.90	35.18
	Dauphin	44	17 " 38.6	27 " 61.4	56.12	34.65
	Northumberland	182	51 " 28.1	131 " 71.9	65.67	43.01
Agr.	Berks	848	380 " 44.8	468 " 55.2	36.36	32.13
	Susquehanna	323	72 " 22.2	251 " 77.8	40.31	25.32
Bit.	Clearfield	472	158 " 33.4	314 " 66.6	42.98	33.43
	Westmoreland	812	301 " 37.0	511 " 63.0	53.97	41.10

teachers are lower. A comparison with two bituminous counties shows a larger percentage of male teachers employed there, while the average salaries paid per month are lower, excepting that paid females in Westmoreland county, which exceeds the general average in the anthracite coal fields by 51 cents per month. These comparisons, as well as that made of the teachers of the State as a whole, show that the percentage of female teachers in our communities is higher than the general average, while the salaries paid them are also higher. The large percentage of female teachers employed in our schools is accounted for by the fact that there are very few other openings for girls to make a living in these coal fields, and the ambitious daughters of mine workers, anxious to secure means to supply their increased wants under a rising standard of living, crowd the teaching profession where good salaries are paid, if compared with salaries earned by young women in stores and offices. Young ladies in these coal fields who have charge of large stores or departments therein get only $8 or $10 a week, of 72 hours' work. Female teachers get $40.59 for about 160 hours' work.

Is it to the best interest of the pupils to have so many female teachers in our schools? In raising the question we readily confess that the average female teacher in our territory is capable and does her work well. Nevertheless, half our school population is male and the growing boy should be brought into contact with qualities which are distinctively possessed by the male and which he is expected to exercise as he enters upon his life work.* The male and the female have characteristics peculiarly their own ; each is complementary to the other. When we meet qualities admirably suited to a female in a male they appear ludicrous if not worthy of disdain. Is it best for the boy in our schools to be daily subjected to characteristics which, if he acquires them, will subject him to ridicule and contempt in life ? When we come to the higher grades, there are other considerations besides the psychological ones which enter into the question. We have some boroughs in our territory where no male teacher is engaged, so that, from the primary department to the time of graduation in the high school, the boy is wholly under female instruction and example. Boys generally graduate when 18 years of age, and the average age of teachers in our counties is 26 years. Girls from 18 to 22 form a large percentage of the teaching force in mining towns and villages, and can any student of human nature affirm that the influences, unconsciously exerted on the lad of 15 or 16 years in public schools under female educators, are best for the boys ? More male teachers ought to be engaged, and whatever obstacle is in the way of their entering the profession and making it their life work, should be removed. The education of our youth ought to be our first consideration and under existing conditions a wrong is done to the growing manhood of our counties.

The selection and appointment of teachers are made by the board of directors, who are elected by public vote. In every borough and township there is a political organization and candidates for the offices of directors of public schools must, as a rule, get the endorsement of the ward or township political leaders before their nomination and election are sure. Politics thus enter into the public school system, and their baneful influence is apparent in many of our towns and villages. Many

* An article of Professor Sanford Bell, M.A., of Mt. Holyoke College, recently published in the *Independent* corroborates this judgment. His conclusion is that for the period of adolescence the more decisive and aggressive influences of male teachers should lead.

scandals due to the greed, nepotism, corruption and misappropriation of school directors have been exposed in our communities, while in almost every borough men seek the office of director in order that their child or relative may have an appointment as a teacher. Instances exist where county superintendents, fully competent to discharge the duties of the office, have been ousted from their position because they refused to be governed by directors in the question of granting certificates of qualification to teach. The superintendent of Luzerne county says in his report of 1901, that a few of the teachers are drones, who are kept in the school because "they have political influence or a friend on the board." The superintendent of the Dunmore schools said in 1900: "There have been many changes in the teaching force during the last two years, mainly on account of politics. Nationality, politics, religion and favoritism should not enter into school affairs, as they always result in injury to the schools." One of the directors of Schuylkill county scored his own class for seeking their own interest and not that of the public in the discharge of their office, and affirmed that boards seek "political influence and prestige rather than the welfare of the rising generation in their town or borough."

The quality of men elected to the important office of school director is far from what it ought to be. Is a man who knows not how to read and write the English language qualified to exercise intelligent supervision over institutions where the youths are taught their mother tongue? What has the fact that a man is a Republican or a Democrat to do with his capacity to be a school director? Many are elected to this office because they are good party men and not because they are qualified to serve the borough by high ideals of what a public school ought to be. There are in our boroughs excellent men who are capable, honest and faithful, but these are generally kept out of office; if a few of them are chosen they are hampered and discredited by the minions of the machine, and their efforts at reform frustrated. When a saloon keeper or a local politician incapable of earning a decent livelihood gets on the school board, he feels that he is fully competent to direct the education of the young and dictate methods to the superintendent or teachers. The electors themselves think not that the success and efficiency of the public school largely depend upon the quality of the school board, and that the best interests of the schools demand that candidates for these honorable offices should themselves have had a good education and have studied the literature of pedagogics. Teachers have a great duty to perform in educating the boards of directors, and working for the elimination of the "spoils system" from our public school directorship. Much good is being done by the Directors' Association, where attention is called to required improvements in the personnel and competency of school boards, and an exchange of ideas between directors and educators is made. These meetings will undoubtedly do much to arouse public attention to the necessary qualifications of school directors, and will ultimately bring forward a better type of men as candidates for the office. In mining towns, however, the first step in the necessary reform is to eliminate both the influence of the local machine and the saloon in the election of school directors. If this is not done, reform is hopeless, and the citizens will have to submit the care of their children's education to agencies that corrupt and debase.

The political type of school director, which unfortunately prevails in these mining communities with rare exception, is detrimental to the selection of teachers upon whom depends the efficiency of our schools. Dr. Schaeffer says that "one cannot help admiring the courage and wisdom which many school boards display in resisting political influences when they select their teachers," but in the anthracite coal fields the Superintendent of Public Instruction would have few instances to kindle his admiration for the type of men he depicts. "Ward politics, church politics, family politics, and sympathy politics" come into play, and the qualifications of the applicants are secondary considerations.

The teachers must obtain certificates of qualification before they are able to teach, but they may teach holding only a provisional certificate, which must be annually renewed. In

A Group of Anglo-Saxon School Children.

A Group of Sclav School Children.

Lackawanna county the number of teachers holding provisional certificates has been reduced from 75 per cent. in 1893 to 28 per cent. in 1901. So that in this county 72 per cent. of the teachers now hold professional certificates. In Luzerne county only 6.1 per cent. hold professional certificates, and in Schuylkill county the percentage is still less, being only 5.6. Young girls, living in the township or borough and holding provisional certificates, secure positions as teachers when others far better qualified are discharged.

One of our county superintendents said in his report of 1901: "It is a matter of regret that as the number of teachers increases in some districts the board begins to make room by dropping from the list those who are non-residents." This tendency of school boards to hire local residents as teachers draws also a protest from the State Superintendent, who affirms in the same year: "To prevent the influx of teachers from rural districts and to reserve lucrative positions for their own people, city authorities are constantly tempted to build around themselves a Chinese wall in the form of local regulations which force the employment of home talent regardless of teaching ability, and lead to results similar to deteriorating effects that follow from constant inbreeding on the farm." In all our boroughs and townships the tendency is to hire local talent and shut out all outside applicants. It is the bane of politics whose cardinal doctrine is the "spoils system."

Under this system corruption is inevitable. Candidates run for the office of director for the sole purpose of securing positions for their relatives. In the town of Shenandoah 66.6 per cent. of the positions in the public schools are held by representatives of a people that does not form 20 per cent. of the population. In the town of Olyphant, a few years ago, the 18 teachers were of the same nationality, which formed only about 33 per cent. of the population. The priest, a public-spirited and broad-minded gentleman, vigorously denounced the injustice from the altar and exerted his influence to right the wrong.

In the matter of the selection of teachers questions of nationality and religion often enter. The borough of Dunmore was for many years the scene of struggle between conflicting creeds in the selection of teachers, and both contestants were equally acrimonious and vindictive. We have boroughs where religious creed is an impassable barrier to an appointment no matter what be the qualifications of the applicant. In this matter both Protestants and Catholics are equally guilty. National bias and religious bigotry are anachronisms. They have no place in our age and generation, and in no department of human life is their baneful influence more deplorable than in our public school system.

There is great need of reform in the personnel of our school boards and the reform can only come by a higher conception by electors of the qualifications of directors. A board that will, each time its political complexion changes, oust 50 per cent. of the 37 teachers engaged by it, for no other reason than "to the victor belongs the spoils," is a curse. A board that will buy a plot of ground for school purposes and pay threefold its market value is a public enemy. Members who will demand bribes from superintendents and teachers before they will vote for their appointment are wholly oblivious of their public duty and devoid of conscience. Janitors and teachers are held up by these corruptionists, who demand a portion of their hard-earned salaries before they will be their "friend." Others will hold up bills for honest work and refuse to pay them until the collector consents to a 5 or a 10 per cent. "shave." They are hawks ever on the watch for prey and they think it no great harm to make something out of their official opportunities at the public expense. From such men our school system suffers. As long as they are elected to office, bribery and corruption will exist. Upon needy teachers and janitors they practice their art and fall to a depth from which it might be thought that the dullest conscience would shrink.

The Boys and Girls in School.

There were enrolled in our schools for the year 1901–1902, 123,384 pupils. The schools of purely mining communities are generally full. There are in these boroughs and townships an average of 202.9 pupils enrolled per 1,000 population. There is considerable variation in the eight counties which will be seen by the following table:

Lackawanna	Co.,		182.1 per 1,000 population.
Luzerne	"		206.1 " " "
Schuylkill	"		210.6 " " "
Dauphin	"		217.1 " " "
Carbon	"		222.1 " " "
Northumberland	"		196 3 " " "
Susquehanna	"	(Forest City)	194.0 " " "
Columbia	"		195.0 " " "
Clearfield	"	(Bituminous)	215.1 " " "
Westmoreland	"	"	208.6 " " "
Berks	"	(Agricultural)	193.0 " " "
Susquehanna	"	"	208.6 " " "

If we compare these counties with two counties in the bituminous coal fields and with two agricultural counties, we do not find much difference.

According to the census of 1900 there was 33.7 per cent. of the population of mining towns from the age of 5 to 20 years. This gives us a population of 134,800 youths to whom educational privileges are offered, and whose usefulness to society largely depends upon the influence exerted upon them by our system of public instruction.

The State law provides that no less than seven months' or 140 days' schooling be given the pupils, and no less than four months of night school can be given. We have seen that the average in mining communities is nine months, and in almost every borough the advantages of night school are offered those who work in the breakers or mines. The superintendent of the public schools of Hazleton, in his report for 1902, states: "For the first time in several years, night schools were asked for and opened." This is an exception. In every borough night schools are opened.

The number of our schools, our teaching force, the appropriations made for school purposes, the length of the school year, and the equipment of the various schools, in the anthracite coal fields are not excelled in any part of the State, outside Philadelphia and Pittsburg, so that whatever defect there is in the educational qualifications of the youths raised in these regions the fault is not the lack of appliances to train them.

One of the directors of Schuylkill county said in 1900: "95 per cent. of all pupils never go above the common grades." The superintendent of Mahanoy City schools said in a recent report that there are two or three girls for every boy in the grammar schools. The superintendent of the Hazleton schools said: "In our schools up to ten years of age, the number of boys and girls is about equal in the classes, but when we examine the classes where pupils are from ten to fourteen years we observe that the girls outnumber the boys about four to one." The superintendent of the Dunmore schools said: "It is a matter of regret that so many of our young children are taken out of school and put to work in factories and mines before the simple rudiments of a good education have been completed." In 1902, the same gentleman reports: "Girls and boys whose ages range from 9 to 12 years, are employed in mills, factories and breakers, and this with the consent of the parents." The superintendent of Lackawanna county said in 1901: "In mining districts the night school reports show that a large number of boys under the age of thirteen years is employed in the coal breakers. These boys are getting only two or three months' instruction in night schools each year." In speaking of the public school opportunities, the superintendent of the Wilkesbarre schools says: "It is very unfortunate too, for many homes are really obliged to make a sacrifice of these opportunities. . . . It is not the wish of the average miner to deprive his children of the advantages of the public school." In every section of the anthracite coal fields the same conditions prevail. The boys of the mine workers as a rule leave school before they enter grammar B, and the girls are very little better. In most boroughs the girls outnumber the boys three to one in the grammar grades, and the same is true in the high schools. In December, 1902, Superintendent Taylor, of Lackawanna county, addressed a circular letter to principals of schools in mining towns. We give the answers of these gentlemen in the tabulated form on page 166. The two last districts are agricultural, which present a striking contrast with the others.

According to the observation of these men the average school life of the boys of mine workers would not be four and one half years, while that of the girls would be a fraction over five years. This means that the boys leave school before they are eleven years of age and the girls before they are twelve. In these boroughs the number of scholars in the high school is small and, of these, very few are mine workers' children. The answers also show that 27 per cent. of the children of school age in these boroughs are out of school, while an investigation of the average age of those working in violation of the State law gave 10.7 years. The boys are generally employed in and around the mines, while the girls are employed in the mills. The following table gives the number of children employed under age in the above-mentioned boroughs:

Age.	Number Employed.
9 Years	133
10 "	284
11 "	254
12 "	221
13 "	85
Total	977

This list is not complete. The labor entailed in compiling a complete list of all the children working in the boroughs would be great, while it is impossible to get the accurate age of children when the system of registration of births is so inexact as at present.

A careful study of the borough of Olyphant, by Principal M. W. Cummings, revealed that 33.4 per cent. of the boys from 8 to 13 years was out of school, and 16.7 per cent. of the girls of that age. Of boys between the ages of 8 and 16 years, 254 or 47.3 per cent. were out of school: there were enrolled in night schools 124, but less than 50 per cent. of these attended regularly. Hence in this borough of 6,100 popula-

Place.	Number of Scholars Enrolled.	Average Miner's Child's School Life.	Percentage of Miners' Children in School.	Average size of the High School Graduating Class.	Percentage of Miners' Children Entering High School.	Do Miners' Children Leave Before They Get the Rudiments of Common Education?	Average Age at Which Children of Miners Begin Work.	Number of Children Working Under Age.	What Reason is Given for Leaving School?
Moosic	641	Boys 4 yrs. Girls 6 "	60	5	19	Yes.	10.7 yrs.	69	Must go to work.
Fell Township	592	Boys 4.5 " Girls 5.5 "	95	They graduate from the breakers and mills.	None.	Nearly all of them; at least 85 per cent.	10.7 "	193	Must help to keep the family.
Old Forge	1,455	4 years.	99	6	1	A large number.	10.1 "	515	Go to breaker or mines.
Dickson	1,141	3 "	85	2.5	2	Most of them.	11.2 "	467	Necessity.
Olyphant	1,027	Boys 5 yrs. Girls 7 "	80	4	2	Yes.	11.05 "	391	Necessity.
Taylor	920	4.5 years.	80	No high school.	One miners' boy in grammar B.	Yes.	10.9 "	304	Various reasons but they cannot be relied on.
Lackawanna Township.	624	From 4–5 years.	97	3.5	4	Many.	10.3 "	194	Going to work.
Archbald	1,099	6 years.	99	7	4	Yes.	10.7 "	270	Must go to work.
Waverly	114	9 "	Farmers' children.	7	All farmers' children.	All receive these.	16 "	None.	
Dalton	162	10 "	Farmers' children.	8	All farmers' children.	All receive these.	16 "	None.	

tion, about 200 boys grow up without having mastered the rudiments of a common school education. The average age of 132 scholars in the night schools of Olyphant was 13 years. Eight of these were girls employed in the silk mill, and the average age at which they began work was 12.1 years ; 124 were boys who began work at an average age of 11.05 years. Mr. Cummings stated that it was constantly necessary to advance pupils from the lower grades to the higher, before they were actually qualified, in order to maintain the proportion of the schools because of the exodus of boys and girls before they complete the course in Grammar B., while in night schools all they attempt to teach are reading, writing and arithmetic. The average scholar in the night school cannot read or write with any proficiency.

The same conditions prevail in all our mining towns and boroughs. A careful computation of the average age at which children left school in the town of Mahanoy City resulted in fixing it at 11.26 years. Of the boys of school age (6 to 13 years) over 14 per cent. were out of school, and over 15 per cent. of the girls. In the ward inhabited by Sclavs in the above town the assessors confessed that it was impossible to get an accurate list of the children of school age. If a correct list was made the percentage of absentees would be higher. The superintendent of Nanticoke schools said it is comparatively easy to enforce the compulsory school law when factories and breakers are idle, but as soon as these begin operations it is impossible. During the six months' strike of the Susquehanna Coal Company in Nanticoke in the winter of 1899, the influx of breaker boys of school age into the public school was so great that the board resolved to open two schools to accommodate them, for to incorporate them in the regular classes would practically destroy the year's work in those classes.

If we take typical schools in mining towns we find that the majority of scholars leave before they enter the grammar grades. The table on the next page gives examples of this.

The four first grades are passed by the average child before

Town.	Total Enrollment.	In Primary Grades.	In Grammar Grades.	In the High School.
Nanticoke.	1,918	1,418 or 73.9%	413 or 21.6%	87 or 4.5%
Mahanoy City.....	2,012	1,567 " 77.8	319 " 15.9	126 " 6.3
Olyphant...........	821	591 " 72.0	205 " 24.9	25 " 3.1
Mt. Carmel.........	1,813	1,477 " 80.4	314 " 17.1	76 " 2.5
Shenandoah.	2,899	2,366 " 81.6	453 " 15.6	80 " 2.8

he is eleven years of age, so that from 50 to 60 per cent. of the children leave school before they enter the grammar grades, while the number of pupils entering the high school does not amount to 4 per cent. of the total number of scholars. If we take the schools of Scranton with its mixed population and where the mine workers only form about 30 per cent. of the population, we find the very same conditions. There are 15,733 scholars enrolled; of these 11,657 or 74.1 per cent. are in departments which correspond to the primary grades in the above table; 3,072 or 19.5 per cent. in the grammar grades, and 1,004 or 6.4 per cent. in the high school. A study of the reports of other cities, where the industrial classes live, such as Newark, N. J., and Springfield, Mass., reveals the same conditions. The children of the industrial classes leave school when they are about 11 years of age, and in this the pupils in mining regions form no exception.*

An investigation into the conditions of schools in small mining camps revealed still more deplorable conditions. Here from 90 to 95 per cent. of the scholars do not advance beyond the primary department. The teacher is handicapped, because the scholars cannot be graded, and hence the children do not advance so rapidly as in larger schools. If we add to this the fact that many teachers in mining camps take little interest in their work and regard it only as a make-shift for something that is better, we can readily see that the complaints of parents in mining camps are not wholly without foundation, who say that "their children learn very little in these schools."

* An investigation into the high schools of the eighteen large cities of Pennsylvania revealed the same conditions as are found in the anthracite regions. See an article by the author in the November (1903) number of the *Pennsylvania School Journal* on the "High Schools of our State."

The privileges of higher education are offered in our cities and most of our boroughs, and by the law of 1899 townships of over 5,000 inhabitants can also secure State aid in the establishment of high schools. But it may be safely said that very few children of the industrial classes enter the high school. In the city of Scranton only 10 per cent. of all scholars in the high school were children of men employed in or around the mines, while the mining population of that city forms about 30 per cent. of the whole. In boroughs of from 6,000 to 20,000 population, the percentage of mine employees' children in high schools would vary from 19 to 35 per cent., while the mining population is from 80 to 95 per cent. Wilkesbarre leads all our schools in the percentage of its scholars in the high school, which is 7.93 per cent.; Scranton has 6.4 per cent.; Dunmore 4.4 per cent.; Carbondale 6.15 per cent.; Mahanoy City 6.3 per cent.; Mt. Carmel 2.98 per cent.; Nanticoke 4.5 per cent.; and Shenandoah 2.8 per cent. Of the scholars entering the high schools not more than 25 per cent. graduate. Taking the mining regions generally the percentage of scholars graduating from the high schools is less than 2 per cent. And this 2 per cent. belongs to families well able to pay for higher education. Adam Smith said it was the duty of the state to strive to educate the poor and leave the education of the rich to itself. If we estimate the money spent in teachers, books and equipments in high schools per capita pupil taught there, we find it is far in excess of the cost per pupil in the primary department, so that our society has gone contrary to the suggestion of the above teacher and philosopher. In the 24 high schools in the anthracite coal fields, the average number of pupils per teacher in 1901 was 32, and the teacher's salary ranges from $75 to $125 a month. Thus the average number of scholars taught by teachers in high schools is about half that taught by those in the grammar and primary grades, while their salary is double that of the latter. It is an interesting fact also that the girls form about 61 per cent. of the scholars in the high schools. One of our superintendents comments on this in his report of 1900: "It is a serious state of affairs when we consider that of the number promoted into the high school each year not 25 per cent. are boys, and not 2 per cent. of the whole number of boys enrolled ever reach that department." In the 24 high schools in our territory in 1901 there were 3,358 scholars, 1,312, or 39 per cent. were male, and 2,046, or 61 per cent. female; 499 graduated that year, of whom 189, or 37.9 per cent. were male, and 310, or 62.1 per cent., were female. In 1902, the percentages of scholars were, male 40 and female 60; those of the graduates were, male 36.8 and female 63.2.

In mining towns night schools are opened. Attendance at night schools depends upon the industrial condition of the people. It is a fact that the more prosperous the times the more children are employed, and the more children taken out of the day schools the larger the attendance at night schools. The superintendent of Wilkesbarre schools said that the number of pupils enrolled in the public schools in 1900 was the same as that of the previous year, while from 1893 to 1899 there was a normal increase. He explains this by saying that it was not due to a static condition of the population but "to the improvement of the times and that children are more generally employed in the mines, factories and mills than they were during the several preceding years (1893–1899)." In the year 1898 the number of pupils in the night schools of Scranton was 1,539, but in 1900 the enrollment was 2,883.

The efforts made to teach children in night schools are not productive of large results. Without exception teachers speak disparagingly of these attempts. In the first month the enrollment is from 10 to 15 per cent. of the number of day pupils, but before the first month is ended 25 per cent. of the enrollment has ceased to attend, and about 50 per cent. only would be the average attendance during the four months, while many of the schools are closed before the four months are ended. In Ashland, 105 were enrolled but the average attendance was only 40. In Carbondale, 171 were enrolled but the superintendent says: "Attendance was very irregular and the result was not satisfactory." In Dunmore, 300 pupils were enrolled but not 50 per cent. of them attended regularly. In Mahanoy

A Coal Breaker—Where Some Boys Graduate.

City, 247 were enrolled with an average attendance of 159. In Mahanoy township, 400 were enrolled and less than 50 per cent. attended regularly. In Mt. Carmel, 257 were enrolled and the average attendance was 40. In Hazleton, 236 were enrolled and the average attendance was 100. The superintendent of Lackawanna county said in 1901: "Nearly all our mining districts have night schools open three or four months in the year. But these schools are a poor substitute for the regular and systematic instruction given in day schools, and in many cases, encourage the people to send boys and girls to work in coal mines, breakers and mills several years sooner than they would if these night schools did not exist."

Young boys who have worked during the day are tired and cannot concentrate their attention on the lessons. The night schools are seldom graded and generally the teachers engaged are young girls who gain their first experience in the art of teaching there. If we add to this that the number of pupils to a teacher is not large, we can well understand that a combination of circumstances makes the night school an institution little calculated to inspire the ardor of the pupils. The sum spent in Scranton for night schools amounts to $5 per capita of scholars enrolled. If the average attendance was 50 per cent. below the enrollment, it would be $10 per scholar for four months or 134 hours' tuition. The day schools cost $18.17 per scholar for 990 hours' tuition. By this comparison the night schools cost about four times as much as the day schools and still the consensus of opinion of those who conduct them is that the results are very unsatisfactory.

There are many children in the parochial schools. The work done by them varies in character. When they come in contact with the public school they feel the stress of competition and follow the pace set by the latter. This is true of the schools of the English-speaking Catholics. There are many Sclav parochial schools which do not compete with public schools, and the children who attend these are far behind those of their age in the public schools. The law requires that instruction in the English branches be given daily to each pupil of school age.

There are Sclav parochial schools where English-speaking teachers are engaged to give instruction in the English language one half of each day, but there are others which do not engage such teachers. The teacher in charge is a Sclav and there is reason to believe that no instruction is given in the English branches. Sclav children who are not taught the language of our country in school, learn it as best they may on the street and in the breaker, and the English they learn is vile and repulsive in the extreme. Considerations of public policy and social well being ought to lead the men in charge of these schools to see that lack of familiarity with the English language will be a great hindrance to the social and industrial advancement of their children. No one will be injured by this policy as the children themselves, and in this the Sclavs stand in their own light.

It would seem also desirable that all parochial schools make an annual report to the State Superintendent of Public Instruction, giving such information as would interest the public and showing that they comply with the State law as to the requirements of teaching the English branches. This information every citizen, either native or foreign born, should be willing to give and the State Superintendent to receive.

The Boys in the Breakers.

One of our superintendents said that the boys in the anthracite coal fields graduate from the breakers and the mines. It is appropriate then to add the breaker as a school where our boys are trained. Letourneau said that the Targui women knew how to read and write in greater numbers than the men. That is the case with those raised in our territory. The girls are better educated than the boys.

In the breakers of the anthracite coal industry there are nearly 18,000 persons employed as slate pickers. The majority of these are boys from the ages of 10 to 14 years. In an investigation conducted in an area where 4,131 persons wholly dependent on the mines lived, we found 64 children employed in and around the mines not 14 years of age. There were

Boys at Work in a Coal Breaker.

24 boys employed in breakers before they were 12 years of age. In other sections of the coal fields the evil of employing children under age in breakers and mines is worse than in our limited area. But if the proportion above mentioned prevails in these coal fields, there are employed in the breakers about 2,400 boys under 12 years of age, and nearly 6,400 boys under 14 years of age working in and around the mines. The tabulated report of superintendents of public schools in Lackawanna county given above, shows how prevalent the evil of child labor is. Improved machinery for cleaning coal has displaced many boys, and it is hoped that a still further improvement and utilization of such machinery will render unnecessary the labor of boys hardly in their teens in these breakers. No industry demands the service of boys whose bone and muscle are not hardened and whose brain has not been developed for continuous and effective thinking. Muscle without intelligence is annually depreciating, being displaced by machinery which does nearly all the rough work. To stunt the body and dull the brains of boys in breakers is to rob them of the mental equipment which is essential to enhance their social worth and enable them to adjust themselves to the requirements of modern life.

The laws of our State relative to child labor are an intricate mass of confusing statutes,* which well illustrate the legislative jobbery of our representatives, who disregard both science and history in their eagerness to do something whereby their politi-

* The law of 1849 provides that no child under 13 years of age can be employed in any factory and all children under 16 years can only be employed for nine months and must attend school for three consecutive months each year. The law of 1899 says that any child between the ages of 13 and 16 years can be employed if he can read and write the English language intelligently. The law of 1849 says that no child can be employed for more than 10 hours each day and the law of 1893 gives permission to employ them for 12 hours a day. An act of 1891 gives permission to employ boys at 12 years in the breakers and boys of 14 years in anthracite mines and an act of 1887 makes it a misdemeanor to employ a child under 12 years of age in mills, factories, mines, etc. A law of 1885 permits the employment of boys 12 years of age in bituminous mines, while boys of 10 years can be employed outside bituminous collieries. In the last legislature (1903) a law was passed raising the age at which boys can be employed in breakers to 14 and in the mines to 16 years, but it is questionable whether this law applies to bituminous as well as anthracite mines.

cal prospects may be enhanced. The law requires every employer to keep a register of all boys employed under 16 years of age which may be seen by the inspectors. No employer does it. Certificates from the parents or guardians of the child, stating its age, are required before the child is employed. Employers secure these but they are not reliable. The employer is protected, the child sacrificed, and a premium is put on perjury.

No industry in the State is so demoralizing and injurious to boys as the anthracite coal industry. For the last half a century these breakers have been filled with boys who should have been in the public schools. They were put to work before they acquired the three "most essential parts of literary education, to read, write and account," and failing to acquire these to the degree in which it is necessary in order to derive pleasure and utility from them in daily life, they grow up in illiteracy, and by the time they are young men many of them cannot read or write their mother tongue. If society in anthracite communities is to be safeguarded against injuries which can be avoided only by increased intelligence, greater attention must be given to the public education of the children.

Necessity often accounts for the presence of boys in the breakers or mines. Many of the advocates of reform lose sight of this. There are many widows and poor families in these coal fields that need the wages earned by these children, and it would be well for kind-hearted people, who consider only the general desirability of fuller education of these boys, to remember this. On the other hand there are many parents who exploit their children. Of the 64 children employed as above referred to, 35 of the parents owned their own homes. Of the nationalities represented the Sclavs were in the lead, but the English, Irish and Welsh followed closely, while 12 of the parents were native born. These parents do not see that a liberal education to the boys is a better investment than to build a house. Solon made a law which acquitted children from maintaining their parents in old age who had neglected to instruct them in some profitable trade or business. Some such law is necessary to-day in anthracite communities to force parents, financially able, to keep their children in school until they graduate from the common branches.

The breaker, where most boys of mine employees begin their life as wage earners, is not favorable to the intellectual development of the lad, however bright his parts may be. Over the chute where the coal passes he stoops and with nimble fingers picks out the impurities. In breakers, where water is not used to wash the coal, the air is laden with coal dust; in winter the little fingers get cold and chap, and at all times when the machinery is in motion the noise from revolving wheels, crushers, screens and the rushing coal is deafening. In such an environment there is nothing to quicken the talent or develop the æsthetic sense of a boy. All is depressing and the wonder is that so many boys who began life under such conditions have been able to rise to prominence in the various spheres of life.

The boy learns many things in the breakers and in the mines. The hard conditions do not dampen the ardor and crush the spirits of the average lad. Most of them are bright, cheerful and full of tricks. They have a good appetite and with dirty hands the contents of the dinner-pail generally disappears. They have their "spats" and fights, and woe betide the man who injures one of them. They are full of fun and frolic, but their curiosity sometimes leads them to injury and death. Many of them fall into the machinery and are mangled, or down the chutes and are smothered. Of all deaths in this risky business the death of one of these boys is the saddest. To witness a funeral procession of a boy hardly in his teens and the cortège made up of his companions in the breaker, is a sight sad enough to melt a heart of stone, and every humane soul asks: "Is this sacrifice of youth necessary for the prosperity of the mining industry?"

There are three things which boys learn in the breakers; they are chewing and smoking tobacco and swearing. Some indeed have learned these before they begin to work in the breaker. Old Abijah Smith said, in his reminiscences of the early days of anthracite mining, that no youth would think of using tobacco before he was 18 years of age. Times have changed in the Wyoming Valley and many lads now contract the habit before they are in their teens, while boys playing on the streets use profane language which horrifies the morally sensitive. Sclav boys when irritated swear shamelessly and afford considerable mirth to their seniors. Many boys trained in a religious home resist the temptations to obscenity and vicious practices so common in and around the mines, but it requires unusually strong moral qualities to develop moral character under conditions so unfavorable.

One of the greatest enemies of these boys is the cigarette. In a mining town where this curse of boyhood was sold in three stores, the consumption was 1,200 boxes or 12,000 cigarettes a month. Miners who smoke use the pipe or a cigar, so that these cigarettes were sold to boys from 8 to 16 years.* There were 480 youths of that age in the borough, so that the consumption per capita was 25 cigarettes, providing all of them smoked. If half the youths — many novices and some veterans — only indulged the per capita consumption per month was double. This evil prevails extensively in mining towns. One of our public school principals was so convinced of the prevalence of the habit among his scholars, that he went to the stores selling cigarettes and asked the traders not to cut the boxes, for many tots came to buy two cigarettes for a penny. The practice of cutting the boxes still goes on. Careful observation of the physical, mental and moral injury wrought by this habit upon boyhood ought to move every community to wage a war of extermination upon this foe which destroys so many boys. Anti-tobacco leagues are sadly needed here. But what hope is there of reforming the boys when the fathers are so addicted to the habit? A superintendent says: "Only one of our teachers uses tobacco; nearly all of the men in our town do use it, ministers, lawyers, doctors, Sabbath school superin-

*The last legislature has made it a crime to sell cigarettes or cigarette-paper to youths under 21 years of age. But of what good are laws unless they are executed? The age limit was 16 years and tots of 8 years purchased cigarettes freely.

tendents, etc. Many of these men stand high in the community. . . . What chance has a poor female teacher that is not considered worth more than $28 per month with her children, who can go out and earn more picking slate than she?"*

There are many other practices among these boys which sap their physical and moral powers. In Lackawanna county a practice known as the "knock down" prevails among the boys. They take regularly from their pay a certain amount before they give their wages to their parents. Some of the coal companies afford the boys an opportunity for this practice, by not issuing a statement of the wages earned by them. Few parents know the rate of wages paid the boys and the time worked by them. They can only find this out by asking the boss — a thing the average parent will not do. Fathers working in the same colliery as their children are so indifferent to the children's earnings, that they know not when the "knock-down" is practiced by the boys. The boys are exceedingly skillful at the business. Many of them live in the same neighborhood and know that their mothers compare notes at pay-day. In order to guard against detection which may arise from a discrepancy in the pays of boys rated alike, they meet and agree to take out the same amount. Boys take in this way from 50 cents to $1 out of their two weeks' pay. In a local strike in 1900, some fathers complained that the boys did not get the regular rate of wages. When shown that they were paid the standard wage the parents were mortified to learn that they were victims of the "knock-down" habit. The revelation occasioned considerable comment and when a company of men discussed the question, one of them said: "It's an old trick: we used to do it ourselves." No one contradicted him, and some fathers practice it still — they hide a bill in the "bacca-box" before they hand the pay over to the wife.

Many of the boys patronize the slot machine, while some of

*Near a barn in one of the alleys in Mahanoy City, two girls, neither of whom was 12 years of age, were seen stealthily smoking cigarettes. While laws are passed in Harrisburg against the sale of cigarettes to minors, the evil grows and shall our young girls fall victims to that which ruins so many of our boys?

them follow with great zeal cock-fighting and stake 5 or 10 or 25 cents on the main. Most of the small boys, however, spend their money in luxuries, and to watch these boys on pay-night in the candy shop is one of the most amusing sights imaginable. They compare their cash; they count their change; they boast how much ice cream, candy, peanuts and soda they consume. The small boy lays away his cigarette very stealthily, while the veteran puffs boldly into the air. The lad of 16 years is about to pass from the candy store, but still lingering where the younger boys are, he feels the dawn of independence, and smokes a cigar to the envy of the smaller lads. All the rivalry, the cunning, the shrewdness, the vanity and the follies of life are seen here as in a microcosm. It is the drama of life in its pleasures, anxieties and pains.

Boys from 12 to 14 years spend from $1 to $2 a month. Those limited to 50 cents or a $1 "blow it in" on pay-night. Those having $1.50 to $2 are "flush" the night after pay, but the evening following they are all on a par — every pocket is empty. The only time the economic vision of these boys is exercised is when the circus comes to town. Then close figuring is done. They come to the last 30 or 25 cents. That they stow away for the expected night, sacrificing the pleasures of the moment for the promise of a good show. Stores which give the boys "tick" soon get out of business. A boy that owes 25 cents steers clear of that bill. The small boy's trade can only be held on a strictly cash basis.

When the lad reaches 16 or 17 years he leaves the candy shop. He feels himself above the small boys that congregate there and he hankers for something other than the "soft stuff" sold in them. It is the turning point in the young man's career. From his early boyhood every pay-night meant a dissipation after the manner of boys. He still craves for that excitement and dissipation and, forsaking the candy store, he finds only one place of welcome — the saloon. Candy is no longer the basis of his dissipation. It is beer and tobacco. When this hour comes many are the boys in mining towns who frequent saloons, for there is no other place provided to meet their requirements. Right here philanthropic efforts should be put forth in anthracite towns and villages. Money taken out of these rich coal deposits cannot find anywhere in the land a better opportunity for good. The founding and endowing of educational and social institutions on a grand scale is become the fashion of rich men of to-day, but it has not begun in these coal fields. For the last half century the sons of anthracite mine workers have been left to the saloon, the dancing hall and the theatre, and lawlessness, irreverence and crime have steadily increased. Is it not time for the leaders in our society to turn their attention to the degeneracy which has gone on apace, and plant institutions for the benefit of these youths whereby they may be helped to better manhood and find that there are higher pleasures in the world than those of sense?

Human nature in the boys of the anthracite coal fields is the same as that of any other crowd of boys. A group of them is equal in original talent to any group of children, but they are planted in hard, coarse soil, and their physical, intellectual and moral parts are stunted. What we need is to give these boys a better environment. Raise the age at which they can begin work to 14 years * as it is in most other States; arouse public sentiment to the rights of children to a liberal education and to the wrongs of greedy parents who perjure themselves and exploit their children; establish a system of public aid whereby widows will not be forced to send their tender boys to the mines or breakers; then possibly boys of tender years can be kept out of breakers. Ignorance will necessarily lead to confusion and industrial crises, which bring disaster to all interests but which inflict greatest injury on the working classes.

* This chapter was written in 1902, previous to the passage of the law raising the age at which boys can begin work in the breaker to 14 years. An inquiry into the effect of this law, made last October, leads us to believe that it is not enforced. When it was signed by the Governor, an effort was made in some localities to enforce it, but boys who were sent home this month returned the next with the new certificate from perjured parents. In some towns more boys of from 11 to 13 years are in school this year than last, but from most of our towns and boroughs the report comes: "No apparent effect that we can see." With a worthless system of gathering birth statistics, with parents who regard their children as productive agents, and with politicians in control of all civic offices, what hope is there of keeping tender boys of 12 and 13 years from the breakers?

Can our Educational System be Improved?

In speaking of the educational needs of Alabama, Professor Roof, of Birmingham, said the greatest need grew out of the lack of funds. In Pennsylvania we have seen that there is no lack of appropriations for school purposes. There is reason to believe that they are so liberal as to lead to corruption of directors and negligence on the part of teachers. If the public teachers were more directly dependent for their salary upon the parents of the children, the home would take greater interest in the child's education and a wholesome influence would soon be exerted on both directors and teachers. Parental responsibility needs quickening, and a larger number of strong and well qualified teachers should be engaged, who make teaching their life work, and whose superior virtue and natural ability would procure from the young people in their care, reverence and respect for authority.

The State has virtually taken all the responsibility for the education of the child from the parents. A free text-book law was passed in 1885, but was not effective till 1893, when authority was given directors to buy "books in all the required branches of study," so that parents have no concern even for books and slates and pencils, and the child is taught from earliest youth to look upon the State as its guardian and the source which gratuitously furnishes all it needs. We have a system of compulsory education and an attendance officer who marshals the children to school, so that parental authority slumbers and children feel that State interference is necessary in the discharge of ordinary duties. The laws of 1899 and 1901 say that all children from 8 to 16 years must attend the public school. Those over 13 years who can read and write the English language can go to work. But here in the anthracite fields are thousands of boys and girls working who are not 13 years of age and who cannot read and write the English language intelligently. What must be the reverence of these boys and girls for State authority when the laws calculated to give them a good start in life, are openly violated and ignored with impunity by operators, parents and factory inspectors?

The effect of compulsory attendance and free text-books upon the schools is not apparent. The economic condition of the people has a greater effect than any other force upon school attendance. During the years of industrial depression from 1893 to 1898 the school attendance in Pennsylvania increased on an average of 3.49 per cent., which was equal to the average annual increase of the population. But from 1898 to 1902, the years of revived industrial activity, the school attendance fell off an average of .73 per annum, notwithstanding a more rapid increase of population due to better times. The superintendents of Wilkesbarre, Nanticoke and Hazleton confirm this conclusion. Each of them testifies that the compulsory education law is not effective when industrial activity offers the boys and girls work in the breakers and in the mills. The law is not effective in the majority of our townships and boroughs to force truant children to school, while its moral effect upon society is to relieve the individual of his obligation. In our territory few attempts have been made to prosecute parents for the non-attendance of their children. In Lackawanna county, County Superintendent Taylor states that the enrollment for 1902 was from 5 to 10 per cent. less than that of 1901. His explanation is "the failure of school directors to enforce the attendance law." The children of poor and thriftless parents are the truants and any attempt to impose a fine on them is defeated by their poverty.

What to do with the children of illiterate and greedy parents is a problem which perplexes and confounds our educators. Listen to some of their complaints: "The great hindrance to the prosperity of our schools was due to the irregular attendance of pupils who, through the indifference of their parents, were out of school nearly half of the term." "Irregular attendance is the greatest hindrance to our progress." "Many of our foreign born parents care very little what becomes of their children, provided they are not bothered with them any more than they can help." "Because of indecent homes, little people come to us with a vicious knowledge beyond their years." Among these are many of the Sclavs and the lower stratum of the English-speaking peoples. Then in many of our towns, the Sclav children attend parochial schools which do not give them the training they ought to receive. In one of our towns these children, when at the age of 12 they enter the public school, are found to be three years' work behind children of the same age who have been trained in the schools of the State. When from 25 to 30 per cent. of the children of some mining towns attends schools of this character, the public school, which is the chief factor in the process of assimilation of foreign material, is seriously handicapped in its work.

We have also seen that the school life of the children of mine employees does not extend over 4 or 5 years. Means should be devised to extend this. One way is to establish kindergarten schools where the children of from three to six years can be prepared for entrance upon the public school. School boards have the power to establish these, but it requires an enlightened public sentiment before they can see the benefit of such institutions to the children of the working classes. One of the superintendents remarks in his report of 1902: "Again we need to learn the lesson from the parents bringing their little ones at 4 years of age and trying to get them into school by deception."

Something also should be done for the boys and girls who now leave school before they are 11 or 12 years of age. There are in our territory over 130,000 children of school age, 98 per cent. of whom never enter the high school, and only 15 per cent. enter the grammar grades. Why is it the children do not avail themselves of the opportunities for a liberal education so freely offered them? It is not poverty. That is the cause in a small percentage. The real cause lies in the conviction of parents that our higher education is useless to the average boy and girl. A child of 12 or 13 years ought to have command of the rudiments of a common education. At this age most of our boys grow tired of theoretic knowledge. A smaller percentage of boys than even enters the high school is destined by nature to become eminent scholars. Their eyes and hands were destined for something more than reading and writing. They hanker for something to do, and if opportunity were offered, many of them would be as expert in the use of hammer, saw, compass and chisel, as they are now in the handling of cigars, cards, pipes, tumblers and balls. These practical people see little use to keep the boy in school till 16 or 17 years of age and then send him to the mines. Our grammar and high school curricula are well calculated to turn out accountants and teachers, and there are 30 applicants for every such position. Girls start in stores for $1.50 a week and are considered fortunate if they advance after years of service to $5 and $6 a week. In factories they only average $3 a week, and at this low rate there is no room for all the applicants. Is it a wonder that parents say: "What's the use of sending them to school to graduate?" Some of the most hopeless cases in mining towns are young men who have graduated from the high school and who have nothing to do. They know too much "to work" they say and they become idlers and some of them criminals.

Some will say why do not the boys and girls get out? That is what many of them do, but think of the danger a young lad or girl of 17 or 18 years incurs in going to the city without knowing anything definitely. They may be graduates of the high school, but there are hundreds of these in the city before they enter, and desirable positions are not for the new comers. Scores of boys and girls get out annually and get back also discouraged and embittered. The boys rail against social conditions under which they are doomed to toil in the mines where a few generations ago society compelled its serfs and criminals to work. Would it not be better for the boys and girls as well as for society, if they were taught some manual labor or service whereby they could earn a competency when they leave home? Right here in the coal fields there is annually a great waste of talent. A better system of education would economize it. If Erasmus were made a musician and Mozart a physician, society would have lost much, and yet that is what we do by not affording the children of these mine workers the opportunity of acquiring a trade for which they are best adapted. When custom or circumstances arbitrarily chalk out the field of labor for these boys and girls regardless of their inherent intellectual energies, the peace and progress of society are menaced. Opportunities and facilities ought to be offered them for exercising their faculties in order to lead them to the greatest possible social worth and add to the sum total of human happiness. Establish trade schools for the boys and schools of domestic arts for the girls, and from the coal fields there will emanate a body of young men and women whose useful lives will be a source of joy to any city, and the homes of our people will be filled with greater peace and happiness. Well has Mr. Henderson asked: "If the state teaches physicians, lawyers, dentists and pharmacists, why not smiths, weavers, decorators and machinists?" Ignorance and discontent lead to confusion; knowledge and social utility bring order and stability.

The moral instruction of our youths should also be attended to. General Gobin in his testimony before the Anthracite Coal Strike Commission stated that the greatest wrong of the strike of 1902 was the lawlessness taught the youths of these regions. That was a great evil, but the tendency of our youths to lawlessness is not an accident; it is the result of gross neglect to teach them the principles of righteousness. The evil effect of substituting the State for the parent in the business of education is more apparent in the moral than in any other sphere. The parents have abdicated their position as teachers of everything. Teaching as a parental function has practically disappeared. It was the custom for the family to teach the children the lessons of right and wrong, but it is so no more. This is not done by the public school, and the church cannot effectively do it by a few hours' teaching once a week, while there are many who never come within its influence. Is it astonishing that under such circumstances youths grow up without an intelligent sense of right and wrong? The results are ingratitude to parents, disregard to the rights of men and of property, insubordination to superiors, lack of respect for women, indulgence in sensuous enjoyments, lack of truthfulness and honor, a contempt for civil authority that presages anarchy and loss of faith in all save material realities. Society must suffer because of this, and it can only be remedied by establishing an efficient machinery for the instruction of the young in righteousness.

The reports of public school superintendents in our territory for the year 1901, covering 1,172 schools in the anthracite coal fields, showed that in only 614 or 52 per cent. was the Bible read, while other reports covering 4,969 schools in other parts of the State showed that the Bible was read in 4,205 or 84 per cent. of them. As far as our public schools are concerned no systematic instruction is given in right and wrong, and the mass of our youths is suffering moral atrophy. Their moral faculties are undeveloped and our system of instruction assumes that moral conduct will take care of itself if only the common branches are taught. Need we wonder that harmless people are assaulted, women insulted, peaceful citizens maligned on the street, and our politics are corrupt?

We need a better machinery to train the conscience. Right and wrong in the affairs of conduct should be matters of instruction as well as reading and arithmetic. Righteousness is essential to a people's existence and it is contrary to reason and practice to imagine that the youth will, without instruction, be able to determine between right and wrong. Practical ethics should be taught the youth as any other branch of knowledge. Every intelligent man feels that the need is great, and bigotry or sentiment ought to give way to the exegencies of the situation. Hitherto any attempt at moral teaching in public schools has threatened serious division in the communities. The children and society are the sufferers, and it would seem reasonable that the enlightened citizens of each community ought to adopt some system of moral instruction acceptable to the general harmonies of individual consciences and religious beliefs. This also should be supplemented by the strenuous efforts of all persons who see the need and believe that means ought to be devised properly to instruct the sense of right and wrong in the youth of the land.

CHAPTER VII.

THE INTELLECTUAL AND RELIGIOUS LIFE.

1. What do Our People Read? 2. Club Life in Mining Towns. 3. The Work of Temperance Reform. 4. Our Churches and the Clergy.

What do Our People Read?

In the year 1899 there were 87 papers of all kinds published in the anthracite coal fields, while in 1902 there were 130, which is an increase of nearly 50 per cent., while the increase in population during that period was only 25 per cent. In this time the number of dailies increased from 18 to 26, and the number of weeklies from 65 to 99, which is an increase of 44.4 and 52.5 per cent. respectively. The semi-weekly and the monthly issues remained the same. The circulation as given by the dailies in 1889 was 48,476 and in 1902, 133,997. Thus twelve years ago one daily paper was issued to every ten persons in our territory, while last year one was issued for every five persons in the same area. In addition to this, there were 97,658 copies of the weeklies issued in 1889, which was one copy for every five persons ; and in 1902, 153,743 copies were distributed, which was one copy to every 4.5 persons. Thus, as far as reliance can be placed upon the figures given as to circulation, the dailies have increased in number about 50 per cent. and in circulation about 100 per cent., while the weeklies have increased in number and circulation about equally.

The following classification is made of the political affiliation of the papers which expressed a preference :

	1889.	1902.
Republican	24	38
Democrat	10	17
Independent	30	33
Independent Democrat	4	4
Independent Republican	0	1
Silver Democrat	0	1

About seven of the dailies issue a Sunday edition. If we consider the purpose for which the papers exist we have the following classification :

	1889.	1902.
Prohibition	1	0
Technical	3	3
Law	1	1
Literature	1	2
Labor	1	3
Religious	1	4
Railroad	0	1
General news	79	116

In 1889, only one of these papers was published in the interests of the Sclavs, a sheet edited by some of the priests of the Ruthenians. To-day we have three printed in Polish, three in Lithuanian, three in Slovak, two in Ruthenian and five in German, making all told sixteen weeklies published in foreign languages in our territory. In addition to this, newspapers, magazines, etc., from other cities are sold in our towns. To have some idea of the number and character of these papers which circulate in the anthracite coal fields, we will take two typical mining towns, one in the Northern and one in the Southern coal fields.

In the town of Mahanoy City, with a population of 13,500, there are two dailies and one weekly published. Besides these 750 copies of the Philadelphia and New York dailies are sold, and about 1,200 copies of Sunday editions distributed each Lord's Day. Of the dailies the following is their order in numerical importance. *North American*, Philadelphia *Inquirer*, Philadelphia *Record*, while about 20 copies of the Philadelphia *Ledger*, and the same number of the New York *Times* are sold. On Sunday the following is the classification according to the numbers distributed: Philadelphia *Inquirer*, *North American*, Philadelphia *Record*, and Sunday *Journal*. Thus in this town the Philadelphia and New York dailies circulate at the rate of one copy to 18 persons, while the Sunday distribution is one to 11.2 persons. Besides this about 50 copies of monthly magazines are sold. They are *Munsey's*, *McClure's*, *Everybody's*, *Leslie's* and the *Strand*.

In Olyphant and the second ward of Blakely, both of which virtually form one town having 7,700 inhabitants, we have the following distribution of newspapers. No daily is published here but two weeklies manage to subsist. Scranton with its four dailies is only six miles away, and of these 300 copies are daily distributed. Then about 280 copies of the Philadelphia and New York dailies are sold. Of the latter class, the *North American* leads, then come the Philadelphia *Record,* the New York *Journal* and the New York *World.* On Sunday about 650 papers are sold. The largest circulation would be that of the Elmira *Telegram,* the *Scrantonian* and the Scranton *Republican.* Then come the following in numerical importance: the New York *World,* New York *Journal,* New York *Herald,* Philadelphia *Press,* Philadelphia *Inquirer, North American* and Philadelphia *Record.* The Pennsylvania *Grit* is also taken by a few. Thus of the dailies of all kind, one copy is sold for every 13 persons, and of the weeklies one copy to every 10 persons in this area. Of magazines few are sold in this community.

The Sunday editions are brought into these communities by special trains running every Sunday morning from New York and Philadelphia, solely to accommodate the patronage for Sunday papers. In each depot a package is left, and distributors scatter them throughout the mining towns and patches, the remotest of them being supplied with the marvelous compilation of news, gossip, slander, etc., which is highly indicative of the ingenuity and energy of the publishers and distributors, and which interest and, to some degree, edify the purchasers.

If we compare these two mining towns, the daily circulation of both local and city papers in Mahanoy City is about one copy to every 10 persons, while in the Olyphant district it is one to every 13 persons. Sunday editions supply one copy to every 11 persons in the former and one to every ten in the latter territory. Based upon this careful computation of the purchase of Sunday issues in typical mining towns, over 40,000 copies of Sunday newspapers would be regularly purchased by mine workers, which is a weekly expenditure of $2,000 or

$104,000 per annum. To calculate the expenditure in dailies is more difficult. After careful computation of the money spent in the above towns, our estimate is one tenth cent daily per capita of population, or about $130,000 per annum. Thus above $234,000 are spent in the anthracite coal fields annually in daily and weekly newspapers. This calculation includes the Sclav as well as the English-speaking workers. If the Sclavs are left out of the count, the above sum is spent by the 300,000 English-speaking people directly dependent on the mining industry.

Little else is spent by our people in literature. After long acquaintance with their homes and personal visitation to hundreds of them, we are convinced that practically all the reading done is of the above-mentioned newspapers. The change in population in the last twenty years has influenced the circulation of the weeklies and dailies. In Shenandoah, only half the number of Sunday papers are sold now as compared with the sales of eight years ago.* But, notwithstanding the presence of the Sclav, the circulation of the above-mentioned journals has greatly increased in the last decade. Political friction and rivalry has something to do with this increase in circulation, but a greater factor is the enterprise of the management and the aggressiveness of the local agents and scribes, who eke out a subsistence in journalism.

The character of these newspapers is varied. All the dailies printed in our territory give from one third to one half of their news-column space to gossip, while the weeklies give over one half to this. Attention is given to the movements of the people, and the prurient desire of each patron who is anxious to see his name in the paper is gratified while the tattlers find interesting material for the discussion of their neighbors' affairs. Some of our local journals flourish by ferreting out the current slanders, while the sanctity of the home and the good name of highly-honored citizens are ruthlessly assailed. One

* In Shenandoah, where 1,200 Sunday papers are sold each Sabbath, the Sclavs, who form 60 per cent. of the population, buy only 25 copies. A larger number patronizes the daily paper of the town, but, generally speaking, they do not buy newspapers published in the English language.

of the most respectable dailies of northeastern Pennsylvania commented as follows in the year 1901 upon one of these pests: "It represents about all that is vicious in journalism and it is a severe reflection upon the common sense and common decency of this community that a publication conducted as it has been conducted could acquire support to cause the mischief which it has."

The character of the daily and weekly metropolitan papers are known to all who are familiar with the dailies of Philadelphia and New York. Newspapers, which are careful as to the quality and veracity of the news printed and aim at cultivating sober thought and respect for established institutions, find little or no patronage in our communities. But those which serve up the last sensation in lurid colors and furnish cuts of the actors in life's tragedy, find a ready sale. Our people care less for sober truth than for sensationalism, and the newspaper which deals a pleasing fantasy or indulges in wildest rhapsody is eagerly purchased, while that which aims at being a vehicle of ideas and a guide to sane thought is seldom asked for.

The influential papers in our territory are owned or controlled by men of political aspirations, who are interested in winning campaigns. These are of positive political convictions and are vigorous defenders of the party with which they are identified. The papers which are owned by individuals not especially interested in politics have no well-defined political creed, but are in the market "for business," no matter who seeks their aid. Many of our dailies and most of our weeklies would soon pass out of existence, if the revenue derived from political campaigns and favors shown by successful candidates were cut off. All candidates, however honorable, regard newspaper advertisement as legitimate expense, and the papers are eager for this patronage. Men of culture and wide experience believe it ethically justifiable to pay a sum of money for the support of a newspaper in a political campaign. Each campaign means two contests; the one between rival candidates of the same party for the nomination, and the other between rival candidates of different parties for the office. In each of these the press takes a leading part. Most of the papers are "ready for business," and the independency of most of them means that the candidate willing to "do business" with them has control of the sheet.

It is difficult to estimate the political influence of these dailies and weeklies, but they are a terror to the average candidate for office. A swarm of gad-flies following a roadster is not more persistent or pestiferous. No wonder the politicians of rival parties club together and effect a compromise in order to escape the fleecing of the press. Candidates invariably do business through committees, which take charge of campaigns and regulate advertising material.

Few of the 114 newspapers make money. They barely subsist. But the financial stress under which most of them labor destroys their influence either as leaders of public opinion or advocates of reform in municipal government. Unworthy men are commended for office, and the actions of friends in power are defended, no matter how villainous they may be, while those of their opponents are condemned, no matter how meritorious they are. This extreme partisanship, controlled by the financial interests of the proprietor, is soon scented by the public, and although it still patronizes the sheet its influence in moulding opinion is lost.

Measured by whatever standard one chooses, the conviction still remains, that the major part of the $234,000 annually spent by our mine workers for reading material, is either wasted on trash or spent for that which debases and defiles. Science and art are traduced in order to pander to the depraved sensuous taste of the masses which are annually more and more confirmed in their vitiated tendencies because of the stuff they eagerly devour. The quest for knowledge, which every healthy soul ought to acquire to assist him in his special calling and make him wiser and more prudent, is not the motive of the purchasers of most of these papers. It is rather a morbid delight in reading repulsive scenes in life's tragedy, and in seeing nasty and miserable pictures taken from miry depths.

The modern newspaper can be of positive, neutral or negative value. It can only be of positive utility when it educates the conscience and stimulates moral activity. When it stoops to depraved tastes, or debases by publishing the unseemly, or misleads, then it is of negative value. The influence of the press in our territory is not to raise the level of public morality. More conscience is sadly needed in the journals which enter the homes of our people. There is place here for intelligent and honorable journalism which will help forward all good causes to victory. Before this is possible, however, the impoverished and stunted life of the masses must be made richer and broader; it must be elevated to an appreciation of higher things than sensuous pleasures, vulgar enjoyments and the latest sensation.

Club Life in Mining Towns.

In no department of human activity is the tendency denominated by Professor Giddings, "consciousness of kind," more pronounced than in the organizations of clubs in our towns and villages. We have all kinds: political, social, musical, dancing, culture, citizens', etc. Men join these according to their predilection, and the bond of union is common interest in the same objects and enjoyments. It is impossible to give the number of all the clubs in the anthracite coal fields. We will limit ourselves by describing club life in two typical mining towns.

In 1895, a law was passed in Harrisburg granting the board of directors of any school district outside cities of the first and second class, the power to establish a library in connection with the public school, for the use of the citizens in general. A tax of one mill per year on the taxables of the school district can be imposed for the maintenance of such a library. In the year 1898, the State Superintendent of Public Instruction said: "Comparatively few districts have availed themselves of the recent library legislation." Boroughs of from 10,000 to 20,000 inhabitants in these coal fields have established such a library, but boroughs of 6,000 population and under have not. In the cities of Scranton and Wilkesbarre, free public libraries have been founded by the munificence of public-spirited citizens, but outside these cities few free public libraries exist. Public school houses are not desirable places for a public library. It is an institution for the training of children, and it is difficult to interest the young man, who is on the threshold of manhood, in a library located in an institution which he has left. Furthermore, books are not the things the average boy of the mine workers delights in, and in the public school there is hardly room to spare to keep the books, to say nothing of a convenient place to satisfy the desire of boys for games and amusements. Andrew Carnegie offered Carbondale $25,000 to build a public library, providing the town would give the site and appropriate $2,500 annually for its maintenance. A resident of the town offered to donate the site, but the town council would not make the requisite annual appropriation. Hence, as far as public libraries are concerned very few exist, and the experience of those in charge of the few which exist is, that the children of mine workers, as a rule, do not patronize them.

There are, however, in our larger towns, reading circles, whose members are largely made up of the wives and daughters of professional men. Years ago, many intelligent mine workers founded clubs which regularly met for the discussion of philosophical subjects, which widened their range of knowledge and added to their wisdom. Unfortunately, their descendants are not so inclined; these prefer to spend their time and leisure in amusement and frolic; they have little taste for reading and care less than their fathers did for general culture; their imagination has not been trained to see the beauties of literature and many of them left the public school before they acquired a thirst for knowledge. Also many men, who were intellectually active and deeply interested in the moral and intellectual life of their town, have migrated because of the immigration of cheap labor. Our towns feel the want of these sturdy and thrifty men. The descendants of Anglo-Saxons who remain in these coal fields, are not so steady and thoughtful as their fathers. Life is not so serious to them as it was to the early immigrants from the British Isles.

The absence of any appliances whereby young men can improve their time, possibly accounts for the drift away from intellectual activity and self-culture. This want is to be regretted, and is due as much to ignorance of the requirements of the age as to the penury which characterizes immigrants bent on acquiring possession. Ample wealth has been extracted from these coal basins to provide appliances and men for the moral culture of the two last generations. A start has been made by some of the companies, which has in it the promise of better things, but the greatest need is to awaken the people themselves to a sense of their wants, so that they will be ready and willing to spend time and money in self-culture.

In the town of Mahanoy City, (population 13,500) a Chautauqua Circle flourished for fourteen years, having an average of 20 members. It still exists as a reading circle. Two other reading clubs exist, averaging 12 members each. There are, in this town, 5 members of the Book Lovers' Library. There is here also a public library of about 1,500 volumes connected with the public school. In the four reading clubs the majority of the members are females — the wives and daughters of men in the professions or teachers in the public school. From the 12,000 people, who depend upon the mines for their subsistence, not a single member was taken. Of the patrons of the public library very few are children of mine workers. The books are almost wholly taken out by the leisure class which has had a liberal education in the public school.

The same conditions prevail in the town of Olyphant. Here a Chautauqua Circle had a precarious existence for some years, averaging about six members, all of whom were teachers in the public schools or daughters of the leading citizens. During nine years three attempts were made by ambitious and respectable young men to organize and maintain a self-culture club, but each failed after an existence of from three to seven months. One of the chief barriers to success was lack of funds to carry on the work, but in addition to this the organizations lacked capable men at the helm as well as enthusiasm in the members lasting for twelve months in the year. During winter months the club flourished, but as soon as summer came the woods and the sunshine were greater attractions than books and a stifling room. An attempt to organize a branch of the Christian Union Association succeeded for two years, when financial embarrassment again overcame the members and the work ceased. Another attempt was made to organize a suburban bicycle club. This continued in existence for four years when again financial stress proved fatal.

In all these attempts at organizing self-improvement clubs the one great obstacle was lack of funds. The success of the Railroad Y. M. C. A. is largely due to the fact that the railroad officials contribute two thirds of the money for the erection of buildings and equipment. If suitable buildings and equipment were available in the above experiments probably better results would have been obtained. The same has been the experience of other towns in the coal fields. Wherever attempts to organize clubs for self-improvement have been made financial difficulties have proved fatal. It is also true that few of the boys employed in and around the mines were interested in these volunteer organizations. The members were young men engaged in stores, offices, clerical work around the mines or engineers. The boy who most needs help of this nature was not there and he would not join, for his general culture was not wide enough to enable him to see that 75 cents a month spent in the maintenance of a self-improvement club was a better investment than a dollar squandered in the saloon for beer and tobacco. It is also obvious that these organizations will not succeed unless they are headed by competent men. A building and furniture placed at the disposal of young men without a man intellectually and morally strong in charge of it may be a curse and not a blessing. In this work men count more than money.

Some self-improvement clubs are found among the Sclavs. A society in Olyphant among the Little Russians, numbering 80 members, has flourished for many years. It has a library of 300 volumes and each winter stereopticon lectures and entertainments are planned by the executive committee. There was a man at the head of this organization, Rev. J. J. Ardan, who, unfortunately, has left the field because of differences of opinion on ecclesiastical rights. In Nanticoke, a flourishing club is supported by the young Polanders which is headed by a layman of sterling qualities. Organizations of this nature are rare among the Sclavs, however. There are in most of our large towns citizens' clubs organized, which are generally in charge or under the supervision of the priests. The object of the organizations is to qualify Sclav immigrants for the duties of citizenship. The good work accomplished by these clubs has frequently received commendation from the judges of our courts, who have been greatly gratified at the accuracy and ease with which some of these men answer the questions put to them when applying for naturalization.

The Hazleton District has had many anomalies which have been mentioned in the progress of our study, and yet it is here we find the only practical effort made by philanthropic men to meet the requirements of the youth of the anthracite coal fields. In 1893 the "Mining and Mechanical Institute of the Anthracite Coal Region of Pennsylvania" was organized for the purpose of training miners, mechanics, apprentices and those employed in the care and management of machinery, so as to make them of "much greater value to themselves and to those for whom and with whom they work." In 1894 the institute was chartered. In 1902 its present commodious home, in Freeland, was opened. It has a free reading room, a reference library, four recitation rooms, seven instructors and an enrollment of 138 students. It has sent many students to Lehigh University, some of whom have highly distinguished themselves. A nominal fee is charged scholars, but the institute is maintained by the voluntary contributions of public-spirited men at the head of whom stand the Coxes of Drifton, to whose philanthropic enterprise it owes its existence. Institutes of this nature, discretely planted in the anthracite region, would afford opportunities of self-improvement to capable and ambitious mine employees and result in preparing a body of men, whose technical knowledge and larger intelligence, would greatly benefit employers of labor in this industry.

An institution that has done good service in quickening the intellectual activity of hundreds of young men in the anthracite coal fields is the International Correspondence Schools of Scranton. This year (1903) it has enrolled 16,200 students in the anthracite coal fields, 25 per cent. of whom take mining courses, and 75 per cent. is trained in other branches of knowledge, such as drawing, electrical and commercial courses. The number of Sclav young men entering these courses is annually increasing. They form to-day about 5 per cent. of the total mine employees enrolled. The students range from 18 to 40 years of age and represent nearly all the nationalities in the coal fields. But, notwithstanding the large number of students enrolled by the above school and kindred institutions in our mining towns and villages, yet the total number forms only about 15 per cent. of the total number of persons between 18 and 40 years in our territory. Among the students are many persons of bright parts, whose talents deserve better opportunities and a more liberal training.

The dearth in intellectual and moral appliances is great in mining towns, but the moment we pass into the sphere of clubs for amusements there is a superfluity of them. Their chief diversion is dancing, and although there are not wanting in our communities many who heartily agree with Sebastian Brand:

> "Gedank ich aber nun daher,
> Wie der Tanz aus Sund enspringen sei,
> So mark ich, und mir bleibt kein Zwiefel,
> Dass ihn erfunden hat der Teufel,"

yet the practice goes on with few restraints. The priests sometimes interfere. They have been known to enter dancing halls with whip in hand and break up the dance. "Young America" resents this, however, and it is of less frequent occurrence to-day than in former years. Protestant pastors denounce it from the pulpit, but their words have little or no effect; they have not tried the whip, because —.

Dancing circles aim at social differentiation, but occasionally some of the "smart set" forget their social standing in the intoxication of a public dance. Lippert classifies music and dancing with the intoxicants, for the reason that they have a tendency to lull thought and banish care. Whoever has watched those fond of music and dancing readily concur in this view. Young respectable women will so far forget themselves in a dancing hall as freely to associate and dance with men whom they would be ashamed to recognize on the street.

In every town of 3,000 or more population there are from one to eight clubs which regularly meet for dancing. These organizations represent the various grades in the social structure of a mining town. The most select have from two to three dances in a season. The couples are invited and charged about $2.50 for the evening's enjoyment. The committee furnishes hall, music, decoration and refreshments, which represent an outlay of about $200 in a town of 14,000 population. Another club more democratic, but still eclectic, meets for a dance every two weeks. Invitations are sent and each couple pays from 50 to 75 cents, according to the number present and the expenses incurred. No definite program is prepared, and when refreshments are served the charge is $1 a couple. From these gatherings mine employees, performing manual labor, are generally excluded. This club would spend from $25 to $45 on the evening's enjoyment. Then manual workers have their clubs, in which little formality prevails, but which are nevertheless strictly governed by a code of rules which are rigidly enforced. These meet generally once a week. No invitations are issued, no program prepared and no refreshments served. The hall is open to the public; the men pay 25 cents entrance fee and the women nothing. Hats are worn in the hall and smoking is indulged in. Within the hall, the rules of ordinary conventionalities are suspended, and no young man needs hesitate to ask any young woman in the room to dance with him. If a young woman refuses to dance with any man who asks her and, later in the evening, responds to some other man present, the affronted one can complain to the committee in charge and the young woman will be asked to leave. These gatherings are very popular, and, when the summer months come, the club arranges its dances in a public park, so that the diversion goes on during every season of the year. The only time when a preceptible falling off takes place is during Lent and Christmas time. The expense incurred in these gatherings does not exceed $25 in the winter season and about half that in summer.

Besides these dancing clubs, there are a few music clubs. New clubs are organized when a new craze takes hold of young people. A new game or a fad that might become the rage of the season will effect this.

From a careful computation of a limited area of the number of young persons, from 16 to 30 years of age, interested in dancing and games of amusement, we found from 50 to 55 per cent. directly or indirectly attracted to these diversions, while hardly 15 per cent. was interested in technical knowledge or general culture.

From a study of the needs of our young people as manifested in club life, thirst for amusement stands forth most prominently. Music and dancing have ever been the great amusements of all barbaric people, and they still have a hold upon our natures. No curriculum which professes to give men and women accomplishments which fit them for entering or entertaining society, leaves them out. Music by the old philosophers was used as the best means to humanize the mind, to soften the temper and to dispose it for performing all the social or moral duties both of public and private life. Dancing has been equally depended upon to give grace to the movements of the body, expression to the joy of the soul, and response to gladdening influence. But the dancing halls, which are weekly patronized by the working classes of these mining towns, are not fit places for our young, for their minds are not humanized and their bodies are not refined in them. False standards of social life are developed there. Dangerous and daring men have perfect freedom and are under no restraint in cementing friendship with gullible young girls, which often means their ruin. There is no refining influence exerted in the hall and the vices resulting from these dancing halls have frequently been brought home with a rude shock to families in these coal fields.

The thirst for amusements is natural and cannot be suppressed. What we need is suitable places where the young may gather and find a wholesome environment, in which the pent-up activities of the body and the mind may find an outlet, without lowering their standard of morality and laying them open to vice. The bad must be displaced by the good, the debasing by the refining, and the degrading by the uplifting, if the young people are to be protected. In connection with opportunity for pure and healthy forms of amusement and entertainment, those anxious for technical or general literary culture should also be aided. When a better environment is created for our young people and the advantages of increased culture is opened to them, a better moral tone will soon follow and the dangers to which they are now exposed will be eliminated.

The Work of Temperance Reform.

The evil effects and the prevalence of the saloon in anthracite coal communities will be treated in a subsequent chapter; just now, we will consider temperance organizations in our territory as means for moral reformation.

Temperance reform is advocated by churches, special temperance reformers, and by mine-managers from considerations of safety and economy. Of the churches in our territory the Roman Catholics have the most flourishing organizations which are fraternalized into one union. Efforts are also made by Protestant churches, but denominational lines hinder co-operation in a town or village, so that the movement lacks the enthusiasm which comes from federation and numbers. Temperance reformers frequently enter these fields and lecture, but their work lacks continuity. As soon as the speaker leaves, the enthusiasm dies away and the reformed seek the old paths. Labor organizations and operators are also advocates of temperance because of economic considerations, and their power and influence can be exerted to a greater degree than they are in behalf of temperance reform.

In every parish of the Roman Catholic Church throughout these coal fields there is organized a branch of the Father Matthew Society. In addition to this they have the Catholic Total Abstinence and Benevolent Society, whose chief distinction is its insurance feature against accident and death. The same parties are largely members of both organizations. In the Father Matthew organization there are three branches: two for males, the senior and the junior or cadets, and one for females called the Woman's Auxiliary. In the town of Olyphant, where the Irish Catholics number about 2,000 souls, the membership in each of the branches was about 80. In the town of Mahanoy City, where the parish comprises about 5,000, there are about 255 members in the senior Father Matthew Society. In the various parishes of the Roman Catholic Church in the anthracite coal fields there are from 17,000 to 18,000 men, boys and women enrolled as members of temperance organizations. Both branches of the Father Matthew Society are supplementary to each other. The junior is for boys of from 15 to 20 years, and as each comes to majority he is transferred to the senior society. The management of the junior branch is in the hands of the boys, under the supervision of a member of the senior society, while the priest in charge exercises supervision over both organizations. The juniors pay 25 cents a month and the seniors 50 cents. Many of the societies hold real estate while others have money invested in building and loan associations or other means of investment. Some Father Matthew Societies are outside the union, either because of their choice to stand alone or because of friction. Generally, however, the societies affiliate and hold an anniversary when much enthusiasm is engendered by addresses, a parade and a banquet, which greatly advance the cause of temperance. The members of the society accomplish a great amount of good in our towns and villages, and the result is many strong organizations of strictly sober young men, who hold positions of trust and responsibility in the anthracite industry as well as in other spheres of usefulness in these coal fields.

Some priests have also introduced into their parishes branches of the "American League of the Cross." It has three divisions: the first is a pledge of total abstinence; the second is anti-treating, and although the member himself takes his social glass, he promises not to accept or give a treat; and the third is a pledge not to frequent saloons; when the member takes a drink he promises to do so at home. This last division is far less patronized than the two first, which largely confirms the observation of students of the drink problem that men frequent saloons chiefly for social intercourse rather than for drink.

Among the Protestant churches efforts are made to check the evil of intemperance by local organizations of Bands of Hope, Blue Ribbon Leagues, Rolls of Honor, Sons of Temperance Societies, and individual pledges used by pastors in their parishes. These efforts, however, lack coöperation and it is often the case that half a dozen such efforts are carried on in the same town, wholly distinct the one from the other. The only effort at united action is within denominational lines, and this is chiefly limited to discussion and recommendations in the annual conferences of the churches, where, in the nature of the case the question of temperance reform is given a subordinate place. There are some churches doing very good work along temperance lines among their parishioners, while others have no special organization to this end, but insist upon total abstinence as a Christian duty. It is impossible to get statistics of this work for it changes continually. A change of pastor or denominational jealousies may cause the total disruption of a flourishing organization within a year. One cannot escape the conviction, that the work of temperance reform, as carried on by the Protestant churches, lacks organization and united effort. Most of the Protestant churches in mining towns are weak, and separate efforts at temperance reform must ever fail. When local churches learn to coöperate upon a platform acceptable to their leaders some good may come from the movement. But as long as individualistic efforts spend themselves in talks and resolutions and writing pledge cards to children, the evil goes on apace and the young men are impelled to the saloon by social instincts which have no other means of gratification.

The anthracite committee of the Y. M. C. A. is carrying on a work in our mining towns and villages which promises good results. It plants institutions which disregard denominational lines and gives our young people means whereby their passion for social intercourse, entertainment and amusement may be gratified.

The influence of the Miners' Union in our territory has, on the whole, been in favor of temperance reform. The need of reform among the mine workers is great. The Sclavs all drink. An effort by one of their pastors to organize a temperance club failed, for the people did not see the use of it. Public sentiment among them regards temperance agitation pretty much as Anglo-Saxons did fifty years ago. They look upon it as a craze of fanatics, and regard it as uncalled for, unwise and contrary to good social habits. Their saloon keepers are leaders in social and religious life and their business is held in honor. Sclav women as well as men drink a social glass, and no gathering is complete which has not a plentiful supply of whisky and lager on hand. The difference between the Sclav's ethical standard and that of the Anglo-American on this question was well illustrated recently. A missionary church, supported by the Presbyterians among the Sclavs, decided to hold a picnic. On the grounds lager was freely sold to increase the receipts of the day, but the shock given the benevolent Presbyterians who aided the cause proved almost fatal to their generous sentiments. Before temperance reform can take root among the Sclavs, a campaign of education must be instituted. Some of their pastors take pledges from young communicants, who promise to abstain until they are twenty years of age, and in order to do so, the priest urges them to keep away from all marriages, baptisms, dances, balls, etc. But among peoples where these events are of such frequent occurrence, the young people are seldom able to keep such a pledge. One of these pastors said: "You can't rely on them." If the Miners' Union takes a pronounced stand for temperance, and its leaders be men who totally abstain, it will exert a wholesome influence upon the Sclav. Thus far, it has done very little in this respect.

Entrepreneurs in charge of our mines also exert a wholesome influence in temperance reform. Every plant in the anthracite coal fields to-day represents an investment of hundreds of thousands of dollars, and the operators cannot afford to put drinking men in charge of these costly concerns. Many efficient mine officials have been discharged because of drink, and the tendency is to exercise greater discipline in this regard among the miners. Fifty years ago, it was nothing unusual for a miner to take a quart of whisky with him into the mines; to-day no one would attempt such a thing. Many years ago, alcoholic drinks were sold in the company stores, while saloons were run on company premises in close proximity to the collieries. Nothing of this kind is tolerated to-day. Men half drunk are not allowed around the mines and a debauch may cost an official his position. When men, who once drew a salary of $1,800 a year, are reduced to company work at $2 a day because of intemperance, it is a forceful object lesson. Officials of the Delaware and Hudson say that their collieries lose a half-day after each pay because of drink, which amounts to twelve days in the year. Mine foremen have to keep from 7 to 10 per cent. more miners than they need, because of the loss of time by those addicted to drink. There are indications that coal operators will, because of economic considerations, carry on a vigorous temperance reform by eliminating the inveterate drunkards who regularly lose a few days after each pay. Discipline of this nature must result in good, if mine foremen and superintendents will be able to withstand the pleadings and importunity of the drunkard's wife.

Legal efforts in behalf of temperance have been made in our State. High license has been in force since 1887, but this has not succeeded in reducing the number of saloons in our towns. The laws prohibiting the sale of intoxicants to minors and on Sunday are openly violated in every town, while the number of "holes in the wall" is great. Two men sold intoxicants without license on the principal street of Olyphant borough for over a year, and they were only driven out of business when the Citizens' Law and Order League of the City of Scranton prosecuted them. This league has done much good in the mining towns of Lackawanna county in driving into hiding accursed dens of lawlessness, which flourished by the connivance of well-paid officials whose duty it was to execute the laws.

But in our territory the work done by religious, voluntary and temperance organizations, the influence exerted by the Miners' Union and the prosecutions of a league of reformers, do not check the evil of intemperance. These do not go to the root of the evil. They do not recognize the fact that the temperance question is only a part of a far greater economic and social question, which embraces the conditions of life and employment of these thousands of mine employees. In it are involved the question of better homes, greater dissemination of knowledge, education of public opinion and the instruction of youths, and the evolution of a better type of manhood. Until these counts are taken into consideration and work begun along these lines, the evil of intemperance in our towns and cities will go on annually increasing. Every patriotic citizen bewails this great evil, and the time has come, as suggested by Prof. W. O. Atwater in the *Outlook* of last November, that all advocates of temperance reform should find a common platform upon which all could join hands in combating this devourer of the substance of the people. Among the weapons of aggressive warfare should be the suggestions of a body of patriotic citizens of Middletown, Conn., as given by the *Outlook:* "Cheap and clean restaurants, gymnasiums, game-rooms, reading-rooms, libraries, and people's institutes, saloons and clubs where intoxicants are not sold, recreation piers and parlors, and public baths." These in the hands of intelligent and moral men will be of practical service. Arms and the man are needed ere a new epic can be evolved from the lives of the men and women of the anthracite coal fields.

Our Churches and the Clergy.

Religion, in some form or other, has played an important part in the affairs of men from prehistoric times down to the present day, and, notwithstanding the changes which are going on in the forms of religious life, there is no likelihood that man will abandon religion so long as he feels a metaphysical need. A nation cannot cast away its religion as it can its clothes. It is the inner life of the people and permeates its language and poetry, its institutions and customs, its thoughts and ideals. It has often taken fearful forms which no one can fully explain. But there is no doubt as to one beneficent result which these gross forms accomplished; they fixed the yoke of custom on mankind and placed so fearful a sanction upon law that no one dreamed of not conforming to it. And, notwithstanding the gross materialism and rationalism of our day, men cannot break away from the norms of conduct which have religious sanction.

Religion, more than any other socializing force, leaves its stamp on society, but its social value depends upon the degree of development in intelligence and morality of a community. Social progress is realized when reason and judgment displace passion and impulse, and the advance of sound knowledge is fatal to all superstitions. Types of religion based on fear or ceremonies having some assumed remedial effect are fast disappearing, and our only hope of saving the masses from falling into gross superstition is, by offering them a religion of reason which will be a vast positive advantage to the present and future of society. All questions are, at present, submitted to cold intellectual judgments, and the claims of religion are subject to no exception. The church, if it is to preserve its usefulness, must meet the rational demands of mankind. If no nation has ever flourished without a religion, the defenders of the faith must show why it is dangerous for us to abdicate the faith of our fathers. The church, unquestionably, has still its sphere of usefulness in the world. It affords a school of discipline for moral and personal perfection, it binds the conscience and the will to fixed principles, it demands and effects this in the intellectual and moral spheres and thus does a work which no other institution can do and becomes a socializing agency of supreme importance.

There is no dearth of religious organizations in our territory.

Each generation of immigrants to the anthracite coal fields has brought its distinctive forms of worship with it. At no time in the history of the anthracite industry have the mine employees suffered the want of specially consecrated places where religious exercises were conducted. The confluence of nations brought many conflicting creeds and ceremonies which have wrought many moral ruins.

In the last decade, many Protestant churches have been abandoned in these coal fields, because of the migration of many former immigrants and the extinction of foreign tongues which die out as the first generation of aliens passes away. The following list of churches of leading Protestant denominations in the anthracite coal fields shows, however, that we are still blessed with a superfluity of religious edifices.

Denominations.	No. of Churches.	No. of Communicants.	No. of Communicants per Pastor.
Methodist Episcopal	68	18,914	278
Presbyterian	48	13,594	215
Baptist	60	7,752	172
Reformed.	35	9,143	274
United Evangelical	26	2,939	113
Congregational	39	4,979	140
Episcopal	31	5,259	169
Total	307	62,580	Aver. 194

Taking the territory as a whole, there is a Protestant church for every 500 nominal Protestants, and a pastor for an average of 194 communicants. The value of church property held by Protestants in these coal fields varies greatly. The most costly edifice is found in Scranton — the St. Luke's Church — valued at $250,000, while the least costly may be found in mine patches and would hardly be worth $500. Great variation also prevails in the salaries of Protestant pastors. Some clergymen in the cities of Scranton and Wilkesbarre get $4,000 a year, while others in mining towns do not get $400.

The following are the number of churches erected in our territory by the Roman Catholics:

Irish Catholic	63	Greek Catholic	18
German Catholic	10	Lithuanian Catholic	12
Slovak Catholic	15	Italian Catholic	6
Polish Catholic	19		
		Total	143

There are besides these 46 missionary stations and chapels where services are occasionally held. These stations are chiefly supplied by Sclav priests. There are in active service 182 priests and assistants, and, besides these, some missionaries in certain sections of the coal fields working among the foreigners. The nationality of the priests is as follows:

Irish American	105	Slovak	14
German American	10	Italian	5
Polish	19	Lithuanian	11
Greek	18		
		Total	182

There are in the coal fields between 250,000 and 270,000 Roman Catholics. Each parish has an average of about 1,500 souls and each priest has an average of 1,300 souls under his charge.

When the Welsh and English, the Irish and Scotch settled in these coal fields they erected church edifices to worship their God, and the Sclavs do the same. During the last 20 years, imposing church edifices have been erected in our mining towns where Sclavs have settled. The Ruthenians and Poles began to settle in the town of Shenandoah in the year 1880, when the population was about 10,000. In 1890, the population was increased to nearly 15,000 — the increase being made up wholly of Sclav immigrants. To-day, there are over 20,000 people residing in the borough and over 12,000 of them are Sclavs. In 1882 the Ruthenians erected the first Sclav church in Shenandoah, now the Sclavs have five houses of worship in this flourishing mining town which are valued at over $110,000. The German and Irish Catholics have each a church and both parishes have about 3,500 souls, while the Protestants have 12 churches to meet the spiritual needs of 5,000 souls. The invasion of the Sclavs has driven out many English-speaking peoples and some Protestant churches in Shenandoah which flourished 20 years ago are to-day pastorless, and are changed into missionary stations. In Mahanoy City (13,500 population) the Sclavs have five churches, erected in the last 15 years, which represent an investment of about

$80,000. The Irish and German Catholics have 2 churches, while the Protestants have 11 places of worship. In Mt. Carmel (13,000 population), the Sclavs have 3 churches, the German and Irish 2, and the Protestants 14. In Hazleton (14,230 population), the Sclavs have 5 churches, the German and Irish Catholics 2, and the Protestants 14. In Nanticoke (12,000 population), the Sclavs have 4 churches, the Irish Catholics 1 and the Protestants 12. In Olyphant, the Sclavs have 2 churches, the Irish Catholics 1 and the Protestants 7. These are typical mining towns and the Sclav churches erected in them are monuments of the fidelity of these people to their religious organizations.

In five of the above-mentioned towns, whose aggregate population is 66,300, we have 81 churches, which is one to every 818 persons or 120 churches to 100,000 population. In the whole of Pensylvania there are 183 churches to every 100,000 population ; in England 144 to every 100,000 ; while Australia leads the world with 210 to every 100,000. In the above towns the Roman Catholics have one church for every 1496 persons, and the Protestants one for every 460. Thus the former have 67 churches per 100,000 population of their faith ; the latter have 217 to every 100,000 of the Protestant faith. The policy of the Catholic church is to erect an imposing structure and equip it with costly furnishings, so that an average of $25,000 would be put out on their edifices. Most of the Protestants build simple structures which have bare furnishings, representing an average expenditure of $7,000. Thus, notwithstanding the fact that Protestants build about three times as many churches as Roman Catholics do, yet the latter expend about three times as much money on each church as the former do.

The Catholics regard all parishioners as communicants ; the Protestants count those who publicly make profession of faith as members. Of the adult Protestant population about 39 per cent. would be members, two thirds of whom are females. If we take the whole Protestant population then only 26 per cent. would be communicants. In the whole of Pennsylvania the percentage of communicants to the population is 32.8. It is estimated that fully 40 per cent. of so-called Protestants never attend divine services, unless they enter a church when a friend is buried. In a careful canvass of one of our mining towns as to the religious affiliations of the inhabitants, out of 2,900 families visited only one individual was found who discredited the Christian religion and had no sympathy with the work of the churches. The chairman of the committee said we have no "infidels and skeptics." When one of the socialistic speakers, during the strike last year, expressed his contempt of the church and disparaged the Bible, he injured his cause more than aught else he could have done. The mine workers, both Catholics and Protestants, do not keep away from the churches because of lack of faith in the Christian religion. It is rather an indifference to their spiritual interests. Church property in our mining towns and villages represents an investment of from $12 to $14 per capita of population; in all the State it is $16 per capita of population. Nearly $1.75 per capita of population per annum is contributed for religious work, which would be about $9 per family.

The priests of the Roman Catholic faith have charge of an average of about 1,500 souls. The Protestant pastors would only average about one fourth that number. The salary of priests varies from $900 to $1,200 a year, while that of assistants is $600. In addition to the regular salary, the priests get the perquisites which come from marriages, baptisms, burials, catechumenical instruction, colenda, etc. In large congregations, however, when the returns from these sources exceed a certain sum, they are to be turned over to the bishop of the diocese. Parsonages are also provided for the priests; in these the assistants also are accommodated and pay $300 a year for board, etc. The heads of families in the Catholic church are supposed to pay a minimum of 50 cents a month to the organization, while single men pay 25 cents. On Christmas, a special collection is generally taken for the benefit of the priest, to which each adult who can afford it, contributes a dollar or more according to his circumstances. Marriage and baptismal fees range from $5 and over, according to the social standing of the party.

Protestant pastors, in mining towns, get from $300 to $1,500 a year, the average falling below $500. In two of the leading denominations in these coal fields, the average salary of the ministers was $437.57 and $421.96 per annum respectively. We have known men to serve some of these churches for $20 a month. Ministers also receive fees for marriages, baptisms and burials. Parsonages are generally provided for the pastors of the Primitive Methodist church; of the Methodist Episcopal church, and many of the German churches; but this is very rarely done by Presbyterians, Baptists, Congregationalists, etc. Over 50 per cent. of the Protestant pastors in our territory pay rent. Members and non-members in these churches contribute as they please, and seldom is anything said to those who take the privileges offered without paying a cent. A Sclav priest said: "If a man does not contribute his share, we don't want him." A Protestant pastor who denounced the 25-cent-a-month Christians had "notice to quit." Marriage fees among Protestants are $5 and under; baptismal fees, if anything is given, never exceed $1, certificate included; and burial fees are seldom mentioned, for a funeral to a poor family means a ruinous debt.

The multiplicity of churches among Protestants is a hindrance to success. A pastor for an average of 100 families is an inexcusable burden. During the last 20 years, a steady exodus of Anglo-Saxons has taken place from these coal fields. This has affected the Protestant churches far more than those of the Roman Catholics. The former, as their membership diminished, felt themselves in financial straits which soon told upon the quality of their ministers, and drove them to questionable means to raise funds to meet current expenses; the latter, although they lost a fraction of their supporters, were not so much handicapped by financial difficulties, for they had carefully avoided building too many churches. Among Protestant pastors there is little or no coöperation for religious work. Each congregation has all it can do in these mining towns to meet current expenses while all kinds of devices are resorted to to wheedle money out of the public. Just think of the following socials advertised by the supposed fountains of spiritual life in communities: "Klondike social," "a match social," "a guess social," "a bean social," "a fagot social," "a cockle social," "a pie social," "an oyster social," "a tie social," "a marriage social," "a Tom Thumb social," "a mock trial social," "a cracker match," "an April fool social," etc. These are generally held in churches, where divine worship is conducted every Sunday, and the boisterous behavior of young persons, bent upon getting a good time, is fatal to the feeling of reverence, which is the breath of life of true worship. Catholics resort also to unworthy means to secure money for church purposes, but they never permit them to come within the walls of their consecrated edifices. Both branches of the Christian church cry for collection after collection and appeal to the gambling instinct and the craving for amusement in order to keep the institution alive. This cannot inspire joyful veneration, and such means of getting money cannot be to the glory of God and the spiritual edification of the contributors. Socials, bazaars, lotteries, musicals and dancing, card playing, etc., destroy the spiritual power of the church and reduce many of them to the standard of secular clubs. It is comparatively easy for the average member to concentrate his effort on material affairs, but to raise him to that exaltation of spirit which sees "visions of God" is impossible. How far this materialistic type of Christianity is removed from the New Testament type, is known only to such as look with singleness of eye upon those heralds of the cross, who caught the spirit of the Master and were ready to go forth without purse or scrip in His name. These anthracite mining communities need strong men full of spiritual force who are ready for sacrifice and service. Weak Protestant churches also need to learn the lesson of coöperation and amalgamation. Half the Protestant pastors could be well spared from these regions, and one third of the Protestant churches could be put to better use than being the home of a handful of persons, whose chief object is not devotion to the truth but to sectarian tenets, calculated to excite their languid devotion and play upon their passions and credulity.

Fully 96 per cent. of the Sclavs are adherents of the Roman Catholic church. There are a few Protestant organizations among them, but they are weak and far removed from each other. In the Wyoming and Lackawanna valleys, where there are about 40,000 Sclavs, no more than 500 Protestant families are found among them. Sclavs come from countries where Church and State are united, and some of the priests find it difficult to teach their parishioners the system of direct contribution to church purposes. In former years, many of the coal operators collected the monthly contribution to the priest, deducting it from the pay at the office. This custom has now almost ceased — the companies have given up collecting for God, but still continue to do so for Caesar. There is considerable difference between the ritual of the Ruthenians, who belong to the Greek Orthodox church, and the other Sclav nations which are adherents of the Latin church. To avoid conflict, when the question of precedence arises, the bishop of the diocese issues instruction as to the mode of procedure. Sclavs of different nations or tribes, who are forced to worship in the same church, do not dwell in unity and considerable friction is sometimes caused. The relation between the Sclav priests and the Irish Catholic priests and the ecclesiastical authority is often strained. This is not strange. Among the Sclavs, as among every other body of immigrants, the restless and active are found in a much larger proportion than in the communities whence they emigrated. These men, although adherents of the Roman Catholic church, have their distinctive racial characteristics which effect their mode of worship, their forms of government, and their sacred holidays. The difference between a Hungarian congregation and an Irish congregation at their devotions is as great as that between English-speaking Catholics and Protestants. Hence, the Sclavs, when they are subjected to the rule of American-trained superiors, find a considerable difference and many of them are inclined to resent forms to which they have not been habituated.

The Poles have shown considerable independence in their demands for control of their property. The Ruthenians have a degree of freedom in church government not known among other Sclav congregations. Their priests marry; they can call and dismiss a pastor; and the property is vested in a board of trustees. Many aggressive Poles desire like freedom, which has led to many schisms and much litigation.

Most of the Sclavs are devout. They faithfully attend to their devotions and often have to walk many miles to church. The edifice is generally overcrowded, and it is nothing unusual to see the worshipers reverently following the service and devoutly performing the gestures on the side walk or street in front of the church. The cold of winter and the heat of summer make no difference to them; through foul and stormy weather they come to their devotions. Those who come from the patches generally visit the dry-goods store after mass, and every Sunday scores of these men can be seen carrying their purchases home.

Our description of the religious life of these coal fields will not be complete if we say nothing of the superstition which prevails here. It is of the lowest kind, and although it prevails in every part of the coal fields, the type met with in Schuylkill County, where a large number of Germans and Pennsylvania Dutch is found, is grosser than any known in the other counties. It passes under the name of "pow-wowing," and there are in every town "pow-wowers" many, both male and female, who can cure diseases and cast out devils. Marvelous cures have been effected by these men, and never have they failed in cases of burns, erysipelas, nose-bleeding, rheumatism, felons, cuts, eczema, etc. The "pow-wowers" use various cabilistic utterances. Many of them use words of Scripture, while others utter some jargon taken from a mysterious volume known as the "Egyptian Secret." They believe they can kill and make alive "if they have a mind to," and with unerring accuracy they can tell if any witch or evil-eyed person has placed a curse on a child or woman. They give medicine made of herbs gathered at a specific time. Mothers take their children to these traffickers in the black art, but if the child is not baptized in the name of the Trinity the "pow-wower" will do nothing; it must first be baptized. The clergyman and the magician sometimes meet in the sick room —both are sent for to heal the sick; the one prays and the other charms. Some cures are effected without personal contact. All you have to do is to tell the "pow-wower" the name of the sick and the color of his hair, etc., then the magic word is uttered and the sick is whole. In some cases a fearful contest takes place between the magician and the witch, but seldom does the charmer fail to triumph. Patients are generally treated three times, "for there are three persons in the Trinity." The vicarious element comes in often times. The shaman takes the disease unto himself and suffers in behalf of the patient; for this he receives extra compensation either in cash or stimulants to sustain his waning strength. Fees range from 10 cents to 25 cents a treatment. Patients generally pay for they fear the power of the "pow-wower," but some fall into scepticism and never give a cent. Many persons carry charms. Two crossed sticks placed on the back of the neck is an infallible remedy for nose-bleeding. Sacred stones are a sure guarantee against diseases, while an old horse-shoe wrapped up in red flannel and hung over the door brings good luck to the home. Around marriages, births and deaths an innumerable number of signs are looked for which are ingeniously interpreted to bring good or bad luck to the mother, the child or the family.

Here in communities where an abundant supply of churches and educated pastors and physicians reside is found as gross and absurd superstition as in any heathen country. Intelligent and God-fearing clergymen, skillful and well-trained doctors are discredited in the presence of a gibberish muttering "pow-wower" by persons who have been raised in the public schools and in the Christian churches. It prevails not only among immigrants but just as much among the native born. When one of the "pow-wowers" was ridiculed by a clergyman, he turned and asked: "Don't you believe your Bible? Did Christ not say: 'Heal the sick, raise the dead, cleanse the leper, cast out devils?'" and no argument based on the special significance of miracles as to time or place, nor a disparagement of the value of miracles in the realm of truth, would have had any effect upon those present who nodded assent to his position, while there was considerable danger of being classified as an infidel or a scoffer.

The people of these coal fields are in a sense "too religious," for their faith in spiritual realities runs to excess and lands them in gross absurdities. And it seems that pastors and priests have no more urgent duty in these regions than to regulate and clarify these excesses which prevail so extensively in these communities. The religious sentiment, like every other power of man, may fall into a pathological condition and become a source of vicious influence in a community. The therapeutic treatment rests with the church, and the work can only be done by men of culture and strong spiritual insight into the divine order of the universe. Ours are Christian communities and there are many in these towns and villages whose consciences are regulated by intelligent and true comprehension of Christian principles, but among the majority of our workers there is a co-mingling of heathen superstition and reverence for the Christian church, which is mischievous and irrational. This can only be removed by painstaking and careful educational training on the part of the church, which duty it owes to itself and to society. An incongruous, unscientific, disconnected and general presentation of the truths of religion once a week, would not check the gross superstition of the masses, providing all were to attend divine service. A smattering of philosophy and theology, constructed on an erroneous and antiquated foundation, ill prepares a minister or priest to be a guide to lead these people from the by-paths of religious excesses to the sober and sane principles of the Founder of the Christian Religion. A materialistic and realistic press, for seven days in the week, batters their beliefs in spiritual realities, destroys their faith in existing religious institutions and teaches them the gospel of violence, mammonism and mud both by pen and pencil. The means used by the church to combat these vicious aud pernicious tendencies are wholly inadequate, and if it is to retain its hold upon the people and be in the future as in the past, a prime factor in man's moral history, it must throw into an irreligious and materialistic age an army of sons and daughters, well prepared for the systematic and scientific training of the people, in the beautiful and sane principles of the Nazarene.

We feel also that religious ideals expressed in public addresses and taught by systematic training are not enough to meet the needs of the masses. Man has a heart as well as a head; he has eyes as well as ears, and both doors ought to be used to lead him to spiritual realities. What is there in the bare and plain church edifices of Protestants to arouse spiritual meditation? Are not forms and ceremonies of value this day as in days of yore, in order that the people assembled for religious worship may be aided to the higher thoughts and moods which Christianity demands? And have not holy days — festivals when the sainted sons and daughters of past ages are commemorated for their fraternal enthusiasm — their object lessons to the thousands who delve in these mines and breakers? Complaints have been made that the Sclavs have too many holy-days * in the year, for they interfere with the production of wealth; would it not pay even capitalists, if these people on holy-days were more deeply and intelligently to understand the spirit of the men they commemorate? The one thing we fear more than any other in our territory is lawlessness and anarchy, but all are ready to confess that an Augustine, a Benedict, a Xavier, a Francis of Assissi, a Casimir, a Nichols, etc., were not anarchists, and a devout and intelligent commemoration of

* One reason for the multiplicity of holy-days among the Sclavs is, that we have, in the coal fields, representatives of both the Eastern and Western Churches. The Ruthenians and adherents of the Greek Church observe their religious festivals by a different calendar from that followed by the communicants of the Latin Church, hence a colliery, having 200 Greek and 200 Latin Catholics among its 800 employees, will not be able to carry on operations when either group celebrates its sacred festivals. If the spiritual leaders of this people were able to effect uniformity in time in the observance of church festivals a part of the grievance of "too many holidays" would be removed.

them will not produce anarchists. Does not church history teach us that the spirit of the church has also been sane while exerting its regenerating influence upon society? Look at the figure of Ambrose in Milan, Savanarola in Florence, St. Bernard at Clairveau, Thomas a Becket in England, Luther in Germany; Wesley, Finley, Father Matthew and Moody in modern times. The protest against the celebration of the holydays is only a part of the spirit which has well-nigh destroyed all religion from the lives of a large number of our working classes, and this absence of reverence for saints and sacred things leads to disregard for constituted authority and industrial discipline. Can any sane man look with complacency upon the destruction of faith in spiritual realities among the masses? Suppose these men say: "Let us eat and drink, for to-morrow we die," will our commerce and industry, our laws and institutions stand? Study the six months' strike of 1902 in these coal fields. No society will long continue to believe in human law if it loses its faith in God, and the church has work to do in this Twentieth Century such as never before confronted it. No nation that is destined for a long life can dispense with universally recognized ideals, and the moral duties, as presented by the Christian church to every individual, must be the basis of permanent existence to our nation.

How can the working class be led to due appreciation of the vital and spiritual truths of the Founder of the Christian religion? Pure intellectualism will not do it. The eyes and hands should be called into service as well as the head and heart. From long study of the religious need of the men and women of our communities, we are led to the conclusion that it can best be done by restoring art to the churches and exercising the sensuous spiritual powers of men. Protestants, in their fear of idolatry, destroyed the sensuous element in religion, and swung to the other extreme — a process always injurious to society. This movement has had an evil effect upon Roman Catholics themselves. It has vitiated their taste and, in place of the art which adorned the cities of the Fifth Century, we have in these imposing edifices in the coal fields, brick-topping with tin to give the appearance of stone structures, and plaster of paris consoles glued to wooden cornices ; the tin rusts and is full of corroded holes and the plaster of paris falls off, and a miserable sight is presented in many of these churches which are supposed to be temples of the true and eternal God. Some of our greatest social reformers have pleaded for the restoration of art to the people. It is sadly needed and especially sacred art which raises the soul above the world of work and need, of struggle and misery, to that of sacred feeling and passion. Milton said that art is useful "to inbreed and cherish in a great people the seeds of virtue and public civility to allay the perturbations of the mind, and set the affections in right tune." When William Meister visited the Three, to whom he entrusted the care of his child, they told him "one thing there is, however, which no child brings into the world with him ; and yet it is on this one thing that all depends for making man in every point a man." Meister could not say what it was, and the Three said : "Reverence, Reverence." They inculcated it by gestures and symbols, and the visitor as he came to understand the foundation of the instruction imparted to the youth said, "you teach the doctrine to your children in the first place as a sensible sign, then with some symbolic accompaniment attached to it, and at last unfold to them its deepest meaning." And in this method of the Three there may be suggestions to the religious leaders of the present day.

CHAPTER VIII.

THE MEN AT THE BAR.

1. The Three Thousand Saloons. 2. What Does it Cost to Keep Them Going? 3. Are the Mine Workers to be Blamed?

The Number of Saloons.

The saloons are not equally distributed throughout the boroughs and townships of the anthracite coal fields. As above stated, the coal operators who have a monopoly of some villages where their operations are carried on, shut out the saloon. In isolated cases coal companies rent a building for this purpose. But mining camps, having no saloon and located in close proximity to towns where intoxicants are sold, increase the number of saloons in these towns. Thus New Philadelphia has a saloon for every 55 inhabitants, but around it are half a dozen mining patches where no saloons are found. M'Adoo, with 2,200 population, had, in 1901, 31 saloons, or one for every 71 persons; but around it are Honey Brook, Yorktown, Audenreid and Silverton, where no saloons are found. In mining camps, however, some speak-easies are found and the beer wagon also supplies the homes with alcoholic drink. But if the men in "dry" mining towns want to enjoy a social chat over their beer, without apprehension of interference, they must go to the licensed saloons.

The following table gives an idea of the prevalence of saloons in the counties of Lackawanna, Luzerne and Schuylkill for the years specified:

Lackawanna County.

Year.	Population.	Licenses Granted.	Licenses Refused.	Population per License.
1880	89,269	224	14	398
1890	142,088	434	43	327
1900	193,831	591	54	328
1901	199,005	717	29	276
1902	199,005	692	31	288
1903	199,005	719	44	276

LUZERNE COUNTY.

Year.	Population.	Licenses Granted.	Licenses Refused.	Population per License.
1880	166,637	701	138	237
1890	223,567	1,064	182	210
1900	257,121	1,096	134	234
1901	262,712	1,244	40	211
1902	262,712	1,182	119	222
1903	262,712	1,253	81	210

SCHUYLKILL COUNTY.

Year.	Population.	Licenses Granted.	Licenses Refused.	Population per License.
1890	154,163	745	50	207
1900	172,927	1,022	70	169
1901	174,803	1,013	65	172
1902	176,679	1,016	50	173
1903	178,555	1,047		170

The increase in the number of licenses granted in Lackawanna and Luzerne counties in the year 1901 is very marked, while just as clearly a falling off is perceptible in Schuylkill county. In the latter county there was a steady increase in the number of saloons from 1893 to 1897. The maximum number of licenses was issued in the year 1897, when one was granted to every 150 persons in the county, but since then a gradual diminution is observed in the number of licenses issued. In the other two counties no such variation is perceptible. If the advance of 10 per cent. given the miners in 1900 had the effect of increasing the number of saloons in Lackawanna and Luzerne counties, there is no indication of any such effect in Schuylkill county. Of course, the number of saloons in this county previous to 1901 was so great that it is surprising how the proprietors were able to subsist. The effect of the strike of 1902 upon the saloons of Schuylkill county was to close about 50 of them, and the wonder is that more of them were not closed, for one saloon for every 173 inhabitants, means a saloon for every 50 adult males in the county.

The anthracite mining industry is the economic basis of the three counties mentioned above and the hint given by the effect of the last strike upon the saloons of Schuylkill county suggests that the wages earned by mining largely support the institution. The following table shows the number of saloons and the number of tons of coal produced in the three principal counties of the anthracite coal fields.

Year.	Tons Produced.	Licenses.	Number of Tons Produced Per License.
Lackawanna County.			
1891	10,184,347	459	22,188
1900	12,282,108	591	20,781
1901	15,409,040	717	21,491
Luzerne County.			
1891	17,726,559	927	18,043
1900	19,179,573	1,096	17,499
1901	21,396,312	1,244	17,199
Schuylkill County.			
1891	9,957,111	749	13,294
1900	11,606,160	1,022	11,356
1901	13,640,766	1,013	13,465

From the two preceding tables we see that the saloon evil is greater in Schuylkill county than in the other two counties. The chief reason for this is the fact that a larger percentage of Sclavs and Germans reside in this county than in the other two.* These people drink lager as a beverage and regularly patronize the saloons, but although they all drink intoxicants, very few of them, comparatively speaking, are inebriates.

In each of the above counties, however, there are agricultural sections, so that the number of saloons in the county does not give an accurate idea of the place the saloon holds in the social and economic life of purely mining communities. The following table gives the number of saloons, etc., in four towns of Schuylkill county.

* Dr. Hugo Hoppe speaks as follows of the relation between the German element of our population and the frequency of saloons. "Nach einer Erhebung für 1895–1896 waren in der Union 237,235 Schankstellen vorhanden. . . . In erster Linie steht der Staat New York mit 1 : 144 Einwohner, dann kommen Illinois mit 1 : 186, Ohio mit 1 : 214 and Pennsylvania mit 1 : 318 Einwohner, alles 4 Staaten, wo das deutsche Element stark vertreten ist."

Place.	Year.	Population.	Licenses Granted.	Persons Per Saloon.
Minersville.	1890	3,504	24	152.3
"	1899	4,684	44	106.4
"	1900	4,815	50	96.3
"	1901	4,815	47	105.4
"	1902	4,815	47	105.4
"	1903		46	
Pottsville.	1890	14,117	75	188.2
"	1900	15,710	72	218.1
"	1901	15,869	75	211.5
"	1902	15,869	72	220.0
"	1903		74	
Ashland.	1890	7,346	51	144.0
"	1899	6,528	46	141.9
"	1900	6,438	43	149.7
"	1901	6,348	45	141.0
"	1902	6,348	42	151.0
"	1903		41	
M'Adoo.	1897	1,900(?)	39	49(?)
"	1899	2,122	31	68.4
"	1900	2,200	29	76.0
"	1901	2,200	31	71.0
"	1902	2,200	30	73.0
"	1903		32	

Each of the above towns is peculiarly situated. Minersville is a purely mining town. Previous to 1890 it was almost wholly inhabited by English-speaking peoples; now 30 per cent. of the population is Sclav, and, in 1901, 36.1 per cent. of the saloons was in Sclav hands.

Pottsville is the county seat of Schuylkill County and a large number of professional men reside there. Many retired and wealthy families make it their home, and it is also a business center for a large area of mining camps and villages. Hence it represents a mixed population such as is found in the county seat of a mining region.

The coal basins around Ashland are exploited and the mine workers who reside there have to go a distance of from five to seven miles to work. Because of this peculiar condition the non-English elements of our coal fields have been wholly kept out of the town, and the residents are those who settled here in the sixties and the seventies and built themselves homes.

Hence it represents a typical mining town of English-speaking residents.

M'Adoo, as above stated, is surrounded by mining camps where no intoxicants are sold, and those addicted to drink come here for their beer, etc. It is a typical mining town and its population represents various nationalities as found in our regions.

In Minersville and M'Adoo, where the Sclav element is strong, the saloons are most numerous, while they are more numerous in Ashland, where English mine employees live, than in Pottsville, where the professional and commercial classes form a substantial part of the population.

Mahanoy City and Shenandoah are typical mining towns practically wholly dependent upon the mining industry. The following table gives the number of saloons in each:

Place.	Year.	Population.	Licenses Granted.	Persons per Saloon.
Mahanoy City.	1890	11,286	79	143.8
"	1899	13,283	130	102.1
"	1900	13,504	135	100
"	1901	13,725	131	104.7
"	1902	13,725	137	100
"	1903	13,725	143	98
Shenandoah.	1890	15,944	83	192
"	1899	19,884	167	119
"	1900	20,321	166	122.4
"	1901	20,700	128	161.7
"	1902	20,700	177	117
"	1903	20,700	174	119

The reduction in the number of licenses issued to saloons, etc., in Shenandoah in 1901, was the result of a vigorous anti-saloon campaign waged by some of the citizens which was continued in 1902, when a brewer, three wholesale dealers and nine bottlers were deprived of their licenses for a season, because they sold to unlicensed places and disregarded the laws regulating the sale of intoxicants.*

* The influence of politics is more potent than that of reformers. In one of our towns the court refused licenses to five saloons, but a brewer who was "boss" of the town, secured the vote of 10 out of 12 members in the council in favor of granting the licenses. A committee of the councilmen waited on the judge and the licenses were granted.

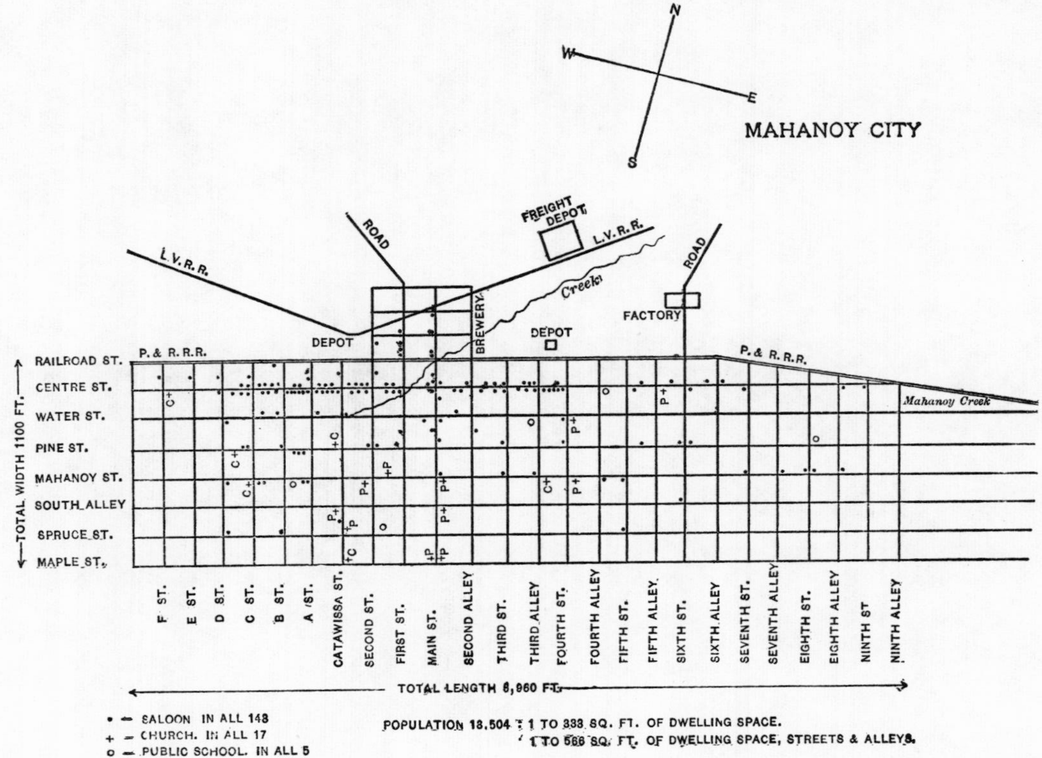
MAHANOY CITY
N
W
E
S
L.V.R.R.
ROAD
FREIGHT DEPOT
L.V.R.R.
Creek
ROAD
FACTORY
DEPOT
BREWERY
DEPOT
P. & R.R.R.
P. & R.R.R.
Mahanoy Creek
TOTAL WIDTH 1100 FT.
RAILROAD ST.
CENTRE ST.
WATER ST.
PINE ST.
MAHANOY ST.
SOUTH ALLEY
SPRUCE ST.
MAPLE ST.
F ST.
E ST.
D ST.
C ST.
B ST.
A ST.
CATAWISSA ST.
SECOND ST.
FIRST ST.
MAIN ST.
SECOND ALLEY
THIRD ST.
THIRD ALLEY
FOURTH ST.
FOURTH ALLEY
FIFTH ST.
FIFTH ALLEY
SIXTH ST.
SIXTH ALLEY
SEVENTH ST.
SEVENTH ALLEY
EIGHTH ST.
EIGHTH ALLEY
NINTH ST
NINTH ALLEY
TOTAL LENGTH 6,960 FT.
• = SALOON IN ALL 148
+ = CHURCH. IN ALL 17
o = PUBLIC SCHOOL. IN ALL 5
POPULATION 18,504 = 1 TO 333 SQ. FT. OF DWELLING SPACE.
1 TO 566 SQ. FT. OF DWELLING SPACE, STREETS & ALLEYS.

Fit Subjects for a Kindergarten. (Sclav children.)

A Portable Saloon Visiting a Mining Patch.

The following table gives the nationality of the vendors of intoxicants in the above towns, which well illustrates the change in the personnel of saloonists in the last decade :

	Mahanoy City.			Shenandoah.		
Nationality.	1890	1900	1901	1890	1900	1901
American.	22	24	24	21	15	16
German.....................	24	24	30	15	13	29
Irish.........................	25	42	40	29	32	20
Welsh.......................	4	7	7	4	3	4
Sclav........................	4	38	33	14	103	94

The Sclavs are annually entering the business in greater numbers. The Court grants very few new licenses to them. They generally buy out some English-speaking saloonist and the license is then transferred. In this way, nearly all the saloons of some wards in our boroughs have passed into their hands. In the first ward of Shenandoah, out of about 50 saloons only three of them are run by English-speaking proprietors, and out of 36 saloons in the first ward of Mahanoy City only six of the proprietors are English-speaking. The law of Pennsylvania is that all vendors of intoxicants must be citizens, so that the Sclav must become naturalized before he can sell intoxicants.

Towns in the Wyoming and Lackawanna valleys present similar conditions if we take into consideration that there are not so many mining camps around these towns where no intoxicants are sold as there are in Schuylkill county. The following table of six mining towns gives an idea of the number of saloons in the Northern coal fields.

Place.	Population.	Licensed 1901.	Places. 1902.	1903.	Persons per License in 1901.
Nanticoke	12,116	68	79	83	178.1
Plymouth	13,649	80	85	82	170.6
Edwardsville	5,665	33	35	33	156.5
Duryea	5,541	37	41	39	150.0
Olyphant	6,100	37	49	49	164.8
Dickson City	4,948	32	42	45	154.6

In all these towns with the exception of Duryea, the saloons depend upon local trade. Duryea is in a condition very much the same as the towns of Mahanoy City and Shenandoah, which supply intoxicants to small towns where saloons are excluded. In Lackawanna township which adjoins Duryea only one licensed place is found to 562 inhabitants. In the year 1902 there was a marked increase in the number of licenses issued to mining towns in the Northern coal fields. The following table for 1902 and 1903 shows this, as well as the marked difference in the number of licenses in various boroughs.

LACKAWANNA COUNTY.

Place.	Number of Licenses.		Persons per License.	
	1902.	1903.	1902.	1903.
Archbald	33	42	163	128
Blakely	5	5	583	583
Carbondale	28	29	483	467
Carbondale Township	9	8	160	180
Dickson City	43	45	112	110
Dunmore	58	70	217	180
Fell Township	28	29	86	83
Jermyn	12	12	214	214
Lackawanna Township	10	14	562	402
Mayfield	13	15	177	153
Moosic	11	13	111	94
Old Forge	52	55	108	102
Olyphant	49	49	126	126
Throop	17	19	129	116
Taylor	30	37	140	114
Winton	31	31	110	110
Total	429	473		

The number of saloons to the population on the Hazelton mountains is about the same as elsewhere in the coal fields, although we find the conditions met with in Schuylkill county intensified here, because of the large number of mining towns owned by individual operators where no licensed places are found. The following table gives the number of saloons to the population in the places mentioned for the year 1901, 1902, 1903. M'Adoo, mentioned above, we insert again for it belongs to this territory.

If we take the whole of this territory, we have here an average of one saloon to every 160.8 persons. Under normal conditions of population three fourths of these would be women and children. Among the Sclavs, however, a large number of

Place.	Population.	Licensed Places.			Persons per License.		
		1901.	1902.	1903.	1901.	1902.	1903.
Hazleton	14,230	71	68	70	200.4	209	203
West Hazleton	2,516	32	32	31	78.6	78	81
Hazel Township	15,143	82	79	74	184.7	191	204
Freeland	5,254	70	69	72	75.0	76	73
M'Adoo	2,122	31	30	32	68.4	70	69

bachelors is found, so we will be nearer the mark by taking two thirds as comprising women and children. By this count, there is a saloon on the Hazleton mountains for every 53.6 adult males.

The number of saloons in towns situated in counties partly controlled by the mine workers is not so great as the above. In Lansford, in Carbon county, there is a license for every 275 persons but Summit Hill and Coaldale are as "wet" as any of our towns; in Forest City, in Susquehanna county, there is a license for every 400 persons. Mt. Carmel, in Northumberland county, is under the curse as badly as any town; it has a license for every 153.2 persons. In every town throughout the coal fields the aggressive Sclavs are found in the business. In Nanticoke over 50 per cent. of the saloons are in their hands, although they only form about 40 per cent. of the population; in Edwardsville the same is true; the same is also true of Mahanoy City and Shenandoah, and it is also true of the towns where intoxicants are sold on the Hazleton mountains. In every instance the percentage of saloons held by Sclavs is in excess of their percentage of the population of these towns. It is possibly due in part to the large number of bachelors found among them, but it is also due to the fact that the only outlet these people have from the routine of daily toil and the monotony of a mining town is the saloon.

The following table of the personnel of saloonists in Olyphant, Lackawanna county, well illustrates the change in the last twenty years:

Year.	Number of Licenses.	Irish.	Welsh.	Americans.	Hebrews.	Sclavs.
1880	8	6	2	0	0	0
1890	33	23	2	1	1	5
1900	37	15	3	1	2	16

We have seen that the labor supply of the future for the anthracite coal industry will largely come from the Sclavs, who build themselves homes and raise families in these communities. If left to the unrestrained influence of the saloon, which to-day plays so prominent a part in their social and economic life, what type of worker will be evolved, upon whom will rest the obligation of faithful and efficient workmanship in this risky business of digging coal? Indeed, we need not limit this question to the Sclav; it applies to the mine workers in general. From the above computations we see that there is in mining towns an average of one license to every 175 inhabitants; or in other words a drinking place for every 58 adult males. President Eliot has said that government regulation is desirable to limit the number of saloons to every 500 or 1,000 inhabitants. Some such regulation is sadly needed here.

In the count we include wholesale as well as retail places, and any one familiar with the wholesale liquor business as conducted in mining towns knows that these houses do more to corrupt the morals and habits of mining camps than retail saloons and restaurants. The same is true of the bottling business. It is nothing unusual to see one of these wagons in a mining camp on pay-day, doing a flourishing business and at which men and boys take their drinks. It is nothing less than a portable saloon which follows the pays, and works an injury which frustrates the laudable attempt of those who aim at shutting out all saloons from certain sections of the coal fields. Mine managers, in general, have justly complained of the drink habit among employees, and unless something more is done than prosecuting speak-easies and violations of the laws regulating Sunday sales; something more than delivering harangues on the evils of intemperance; something more than passing resolutions at public meetings, and signing pledge-cards, the number of saloons will still increase with advanced wages, and by the laws of heredity the rising generation will be more deeply involved in this curse than their parents, and their economic worth and social value will depreciate. For the prosperity of the industry and the well being of our society, to speak nothing of patriotic and religious motives, something practical should be done to reduce the number of saloons in the anthracite coal fields.

WHAT DOES IT COST TO KEEP THEM GOING?

The City of Scranton received $135,000 license money this year, which was equal to $1.32 per capita of population. The boroughs of Dickson and Olyphant received $3,898.44 and $3,554.46, or 79 cents and 58 cents per capita of population, respectively. The money received annually from licenses by our mining towns and villages is probably not far from $400,-000 all of which must come from the wages of the toilers. But this sum represents only a fraction of the money spent annually in the maintenance of saloons. In order to get an idea of what it costs to keep alive this institution, which corrupts our youth, debauches manhood and devours the substance of many families, the following computation will be of interest:

In mining towns saloons rent for from $35 to $60 a month. The outfit necessary to carry on the business amounts to from $500 to $800. A license in mining towns, together with the lawyer's fee, amount to $265. Water rent is about $40 a year. Ice, light and fuel will average $12.75 a month. The wear and tear of pipes, glasses, etc., will average $5 a month. Another $5 a month is spent in inducements on pay-days, such as musicians, dances, free lunch, etc. Saloons are also open from early morning till midnight, and each proprietor must engage the service of a bartender, who generally gets $25 a month with board and lodging. A hired girl must also be kept, who gets from $2 to $3 a week. Thus the average saloonist finds his first year's expenses as follows:

Average rent	$ 480
Average outfit	600
License, etc.	265
Water rent	40
Ice, fuel, light	150
Wear and tear	60
Wages of bartender and domestic	404
Board " " " "	240
Attractions	60
Total	$2,299

The running expenses of the average saloon in mining towns will be the above less the $600 for outfit. On this, however, 10 per cent. interest is expected, and the saloonist and his family must get their living out of the business, which we estimate at $50 a month. Hence the average annual expense in running one of our representative saloons is :

Incidental expenses	$1,768
Interest on capital	60
Wages of saloon keeper	600
Total	$2,428

Thus in order that the representative public house in our towns may clear expenses annually, the proprietor must make a profit of $200 a month on the commodities he sells.

The average monthly business done by four saloons is given in the following table :

Commodities.	English-Speaking.		Sclavs.	
	(1) Sales per Month.	(2) Sales per Month.	(3) Sales per Month.	(4) Sales per Month.
Lager	$210	$195	$400	$140
Ale	130	168	25	0
Porter	14	25	26	10
Liquors	85	35	400	55
Cigars	60	25	50	10
Total	$499	$448	$901	$215

Saloons conducted by English-speaking proprietors in these coal fields are more uniform as to equipment and volume of business than those among Sclavs. The two Sclav houses given above represent the extremes, the English-speaking houses are typical saloons in mining towns. Each of the saloons gave credit, and the first lost in bad debts about $40 a month ; the second, $25 ; the third about $50 and the fourth $15. The profit of each on lager, ale and porter was 100 per cent., while on whisky and wines it amounted to more. One of the men threw a glass of whisky to a cuspidor and said : "There goes three cents"; he charged ten cents for it. Generally speaking, saloonists expect to clear more than 100 per cent. on their sales. The estimated profits of the above gentlemen per month was $245, $230, $500 and $110 respectively.

The first, second and fourth confessed that all they could do was to live, while the third was making money. Number four was able to subsist, for he carried on the business on a small scale in one of the rooms of the house which he owned, and secured the services of his wife and children when he could not attend to the bar. Many of these houses also sell bitters, have lunch counters and gambling devices. The last mentioned are kept by many "to help pay the rent," but the authorities of our counties have in recent years raided saloons having gambling devices, so that they are not so prevalent as once they were and those which remain are kept in back rooms of prominent hotels or transferred to places outside of the city or borough limits.

Saloons doing strictly cash business are rare in mining towns. Practically all of them give "tick" and the man behind the bar must watch his P's and Q's if he expects to stay in business. If a patron contracts a debt to the amount of $10, he is the object of suspicion. The "sticker" is skillful. He drinks heavily and each month pays part of what he owes. When the arrears reach the $50 mark he makes a change and opens an account in another saloon.

Patrons vary greatly in their taste for drinks and in the amount they spend in saloons. The Germans and Sclavs drink lager; the English, ale; the Welsh, ale and porter; the Irish, "what's going," and the inured of all classes "rotten whisky."* The Sclavs are the best patrons of the saloon. They spend on an average more per month on intoxicants than the English-speaking, and they pay their bills better. The average monthly bill of the former is from $4 to $5; that of the latter from $2 to $3. Many persons do not spend more than a dollar a month on intoxicants, while others run a bill of from $10 to $15. In

* Adulterated liquors have been sold in these mining towns for years, but the Government has not thought it its business to interfere. At present (October, 1903), however, there is considerable activity diplayed by the "Dairy and Food Commissioner," and brewers and retail vendors are prosecuted. One of our saloonists complained to a sales agent that his patrons did not like the whisky he last received. The agent went to the barrel, drew a viol from his pocket and poured some of the contents into the barrel, and the whisky was then pronounced by the patrons "bully."

our investigation the highest bill we heard of, contracted in one month, was $45, which the miner paid that same month.

The experience of the average saloonist is that from six to ten men among his patrons must be watched. They are drunkards and he must "shut down on them." Of this class there are more Anglo-Saxons than Sclavs, if counted according to the proportion of each to the population. The saloons do business for seven days in the week; the front door is used for six days and the back door for the seventh. The law of the State prohibits selling on Sunday, but this is the dram-shop's busy day. Saloonists say that the trade is double that of any other day save pay-day. Lang's investigations in Zurich and Hoppe in Berlin show that Sunday is the favorite day for drink. The same may be said of our mining towns. On pay-day mine workers generally meet in saloons. In the Northern coal fields the miner pays the laborer and the saloons serve a useful purpose by accommodating these men with change.* This is considered by many to be a practice which increases intemperance, but we have seen that the drinking habit is more prevalent in the Southern than in the Northern fields, and saloons in Schuylkill county have no occasion to prepare change for the convenience of their patrons. The fact is that mine workers are in the habit of frequenting saloons and, change or no change, they will go there.

The character of the saloon depends upon the man in charge. Some of them are embodiments of all that is vicious and degrading, while others put a little conscience into the business. Some flourish by enticing youths to their infamous dens by allurements which kill both body and soul; others keep "respectable" houses, but scruple not to sell both on Sunday and to minors. One of this latter class was asked what were the means whereby he held his trade? His answer was, "I keep good stuff, give good measure, keep a clean place and sell on tick."

* The coal companies in the Northern coal fields, since last April, pay the miners' laborers at the office, according to the recommendation of the Anthracite Coal Strike Commission.

Charles Gide says: "in the department of the Nord, one drink shop is reckoned for 46 persons, and as out of these 46 inhabitants three fourths are women and children that leaves one public house for every twelve men." Bad as the drinking habit is in mining towns, we find nothing to compare with that. Dipsomania is on the increase, however, for there is little to prevent it. The competition between saloonists is intense and as a class these men are bent on making money, and many of them adopt devilish schemes to swell their ill-gotten gains. Given a body of unscrupulous men bent on money-making and a community exempt of all other means and appliances to meet the people's desire for social amusement and recreation, what is there to stop a development of the taste for strong drink and an impoverishing of the people, so that we will each year approach nearer the Department of the Nord standard?

There are 2,974 licensed places in Lackawanna, Luzerne and Schuylkill counties. In the whole of the anthracite coal fields there are over 3,000. From the above calculation of the income of typical saloons in mining towns we saw it was about $400 a month. If we put down their average sales at $300 a month, which would hardly keep them alive, we will have the sum of $900,000 a month or $10,800,000 a year spent in the saloons of the anthracite coal fields. Of course, this enormous sum is not spent by mine workers only, for as Lawyer C. S. Darrow said before the Coal Strike Commission, lawyers, doctors, ministers, business men, politicians, etc., drink some of the stuff. If one third were consumed by these classes that would leave about $7,200,000 as the share of mine workers, which would be about $4 a month or $48 a year per capita of mine employees. This sum is about one eighth of the wages of mine workers. Hoppe says wage earners spend on an average one tenth of their income, so that the proportion spent by anthracite coal workers is in excess of the average.

Are the Mine Workers to be Blamed?

College, university and social settlements have taught us that wherever concerts and lectures, gymnasiums and amusements, instruction and kindergartens are established, an insensible taste for better things is given the community, and with this comes the desire for better home life, for better dwellings and for amusements of a refined and elevated nature. They have also taught us that the work can only be accomplished by personal influence, and that the working man cannot be lifted above his present surroundings to a more wholesome and healthful influence until public interest in the work is aroused and assistance given all reasonable attempts at reform. We are convinced that this is true of the anthracite coal fields. The saloon can only be taken out of the life of our people by the expulsive power of better things in the hands of capable and moral men, who will arouse public interest in and get financial support for the work. It is absurd to say that the saloon does not meet a social need among our people. An institution that secures the patronage of 80 per cent. of the adult male population of the mining regions is the most popular of all institutions. It is better patronized than the church, the theatre, the dancing class, or technical instruction and general culture. It consumes more of the wealth of the people than our educational system and has a firmer hold upon them than either the press or politics. The attempt made in 1888 to bring the State of Pennsylvania into line with other states of the Union, where the sale of intoxicants is prohibited, was overwhelmingly defeated, and in no part of the commonwealth was the vote for the dram-shop more pronounced than in these coal fields. Temperance sentiment has not increased in the last fourteen years. The saloon is a greater power to-day in the anthracite coal fields than ever before and a Van Dyke bill * will not check its growth or power. It may result in driving the saloon from certain wards in our cities and boroughs, and then wherever the attempt is made we will see the conditions which prevail in Duryea, New Philadelphia, Freeland and M'Adoo multiplied. There are wards in our larger towns that are accursèd plague

* This bill, introduced in the last legislature, was to give wards in cities and boroughs the right to shut out the saloon if the majority of the electors voted to do so.

spots; drive more saloons into these places and their pestiferous and poisonous character will be intensified. The menace to our society does not come from wards where the sober and industrious live and where the saloons are few in number, but from wards where the people are hopelessly given over to the worship of Bacchus. In these quarters the sweeping must be done and what will Van Dyke bills be able to do for them?

We believe in legislative interference where it can do some good, but those who have watched the effect of legislative enactments far in advance of public sentiment, know that they work far more evil than good. Take the laws forbidding the sale of intoxicants on Sunday and election days. They are only observed when a league of respectable citizens gets on the war path, and, as soon as the wave of reform has spent itself, the evil rushes on with greater violence because of the temporary crusade instituted. It is the duty of the Government to put all possible obstacles in the way of its citizens spending their strength in the production and consumption of articles of negative utility, and to encourage them to produce and consume those of positive utility. The more wheat, vegetables and meat the people produce and consume the better they are, and the less whisky, wine and beer they produce and consume, the more intense will be their physical, intellectual and moral life. But direct governmental interference will not effect this. It can only come when the people feel and know that whisky is not as good as bread, and when they will have the moral power to demand the one and reject the other. If we give attention to instructing the people, to providing them with proper appliances to meet their social instinct, to educating the young in self-control and self-respect, legislation will take care of itself. Unless the people are taught self-help and the individual is set right, legislative tinkering will effect nothing.

Loud protestations have been heard from capitalists and entrepreneurs regarding the frightful waste of wages among mine workers who frequent saloons. It is said if they spent less in saloons they would have more in their homes. True. But who are the men who throw stones? We have seen that an average of $4 per month per capita of mine workers is spent in saloons in purely mining towns; that amounts to about 13 cents a day — not a big sum to spend in luxury and it does not go far even in saloons where "all drinks, 5 cents" is advertised. How much do the censors spend in cigars every day; how much in drinks, and how much in festivals at $10 a plate? Lilienfeld has said that society suffers because of hyperemia in one part and anemia in the other. Is not this true of those directly interested in the production of coal? When coal is sold at tide-water for from $4.50 to $5 a ton under normal conditions, of which about $1 is doled out as the mine laborers' share, is it strange that mine workers complain of an unequal distribution of productive wealth, and regard the cry of "waste among the working classes" from the lips of capitalists as the most Pharisaic utterance of the age? In a country where twice as much is spent annually for drinks as for bread, the working classes are not slow to learn that the larger percentage of the drinks goes to the other fellow. Every class of society, whose physical and mental activity is intense, demands relaxation and diversion. Our people cannot afford a trip to the city in midwinter and to the seashore in midsummer; their taste for grand operas and Shakesperean plays has never been cultivated; they know not what the beauties of art and literature mean; is it strange then that they find their chief pleasure in the pipe and the cup? Surely, in the absence of all other sources of pleasure and amusement, the censors ought not to begrudge these people a sip at the fountain of luxury at the rate of 13 cents a day. There is no danger that this rate of luxurious consumption will wreck society. The real danger to our civilization comes from the regal extravagance of the censors, who vulgarly display their riches in mansions, cottages, etc., which the anemic part of society views with invidious eyes and cannot understand why a little more of the good things of life does not come its way.

Goethe said: "I hate luxury, for it destroys the fancy." Cannot we with propriety extend this and say: "I hate luxury for it destroys society?" Jhering has compared the desire for luxury to a steeple-chase in which the vanities of the social classes run. Luxury is contagious. The working classes by the law of imitation follow the pace set by the rich and, under the pressure exerted by a rising standard of living, many sons and daughters of the working classes are made miserable, and society is shaken with periodical cataclasms which are ominous. Every class in the social hierarchy, under the pressure of a rising standard of living, extends the sphere of its felt wants according to its taste, and among a large number of mine workers it takes the form of greater sensual enjoyments. Nitti says: "The lower the economic situation and moral sentiments of the popular classes, the more are they carried to pleasures that are sensual and prolification is more irregular and abundant." The number of saloons in our territory is not the cause of the drinking habit among our people; rather the disposition to drunkenness among the people is the cause for the number of saloons, and the taste for drink grows because of the growing desire for enjoyments which now-a-days carries all social classes to extremes. Where means are limited, as they are always among the working classes, the people are led to luxury at the expense of the necessaries of life, and the effect upon the worker is relaxation of energies and morals and the production of general effeminacy. In this a change must come, either by the voluntary self-abnegation of the leaders of the industrial world, who, to avoid impending ruin, will rise to the altruistic type of individualism whose ideal is "to serve and not to be served," or by the action of economic and moral laws which can ruthlessly tear down the proudest social structure and involve a nation in hopeless disaster.

The vast majority of mine employees are industrious, and although they frequent saloons either for social purposes or to quench their thirst, they neither neglect their homes nor their families. There are many confirmed drunkards in our towns and villages — men dead to all paternal and marital obligations, and slaves of a depraved appetite. They have fallen to the condition of the natives in German colonies, who will only put forth real effort in labor for whisky and tobacco. They are the dissolute and disorderly, the idle and the vicious, who indulge in the use of intoxicants regardless of the distress which their indulgence brings upon their families. They not only spend money in drink but also lose time because of it. These men are found in almost every colliery throughout the coal fields, unless they have fallen so low as to give up working and live wholly by charity. What to do with this class deserves the attention of society. They are the victims partly of their own enervated will and partly of the environment in which they have grown. Society cannot shirk its part of the responsibility and it should provide for these confirmed drunkards who bring so much misery upon themselves and their families. They are wholly unfit to stand alone and, as long as social conditions produce this type of men and women, a special institution should be maintained at public expense, where they could be removed from temptation.

All the Sclavs patronize the saloons. It is nothing unusual to see both women and men of the Sclav races taking their social glass in public houses. The Poles, addicted to drink in their native country, keep up the practice here. They are considered the heaviest drinkers in the coal fields.

In many of these saloons are found clubs where the workmen meet for the discussion of the questions of the day. Bagehot said that common discussion of common actions and common interests becomes the root of change and progress. The saloons in these coal fields afford better opportunity for this than any other institution. The men are shy of a chapel and will not assemble there. The prayer-meeting atmosphere or Sunday-school discipline restricts their freedom and they will not meet where these are wont to be held. They are freer when they can smoke and quaff the bowl. The danger is that men in an atmosphere of dissipation and revelry will not be equal or disposed to the painful effort which is often needed in order to arrive at ideas and principles that are beneficial to both the individual and society. If the atmosphere of a chapel is too restraining, that of a saloon is too relaxative. In the saloon, men are apt to seek that which is momentary in its effect and lose sight of those immutable laws upon which society rests.

Workers in lead mines are insensibly poisoned by daily infusion of small quantities of lead into their system which do not kill at once, but which produce physical disorders and ultimately death. Our men, who are forced to gratify their social instinct in the environment of a saloon, are subject to influences which insidiously bring intellectual and moral disorders. If discussion is to lead to change and progress, an atmosphere must be created more favorable to calm treatment of the economic and moral interests of the working classes. There are deeper needs and more urgent wants than an increase in wages and a larger share in the pleasures and leisure of life, but the saloon, which lives by selling that which gives momentary gratification, will never be the means by which these profounder realities of life will be fathomed by our workers.

Higher wages and shorter hours are not necessarily synonymous with improved economic and social conditions among the working classes. If they are accompanied by a moral and intellectual degeneracy they will only accelerate retrogression. Wages spent in beer rather than bread, in whisky rather than clothes, in vaudeville rather than culture, in exciting romances rather than works of art, will result in deeper misery and inefficiency among the wage earners. If greater leisure is spent in saloons, or on the street corners, or in cock-fighting, or at the slot-machine, or at cards, what will it benefit the workers? Higher wages and shorter hours will benefit our people only when opportunities of culture and amusement will be given them, far removed from the influence of saloons, and when a degree of intelligence and self-respect among the workers themselves will convince them that the thirst for intoxicants is one of the greatest enemies to social and economic well being. Well may the words of the prophet of Chelsea Hill be addressed to our people: "No man oppresses thee, O free and independent franchiser, but does not this pewter pot oppress thee? No son of Adam can bid thee come and go, but this absurb pot of heavy-wet can and does. Thou art the thrall not of Cedric the Saxon, but of thy own brutal appetites and this accursèd dish of liquor. And thou pratest of thy 'liberty' thou entire blockhead."

CHAPTER IX.

THOSE WHO SAVE MONEY.

1. The Banks and the Cash Deposits. 2. The "Building and Loan." 3. Organizations for Sick Benefits. 4. Insurance Against Mine Accidents.

The Banks and the Cash Deposits.

An entrepreneur of wide experience once said: "The supreme folly of a strike is shown by the fact that there is seldom or ever a rich workman at the head of it." In the strikes of 1900 and 1902 in the anthracite coal fields, the men who suffered most were the thrifty and industrious, and with rare exceptions they deprecated the conflict in which they were engaged. A German forced to draw on his bank account, was heard to say each time: "There's $10 thrown away." The man who has a bank account is more conservative than the shiftless who has nothing, and one would reasonably expect employers to look with favor upon their employees who try to save and offer them all possible appliances to do so. Bankers in these coal fields testify that, during the last strikes, the mine working depositors drew their savings with great reluctance, and invariably it was done under stress of want. The truth is that a bank account is a guarantee to the employer of good conduct, steady service and conservative action in time of industrial friction. Some workingmen also feel that they must secrete their savings where their employer will know nothing of them, for they think that their wages will be reduced if the employer finds out they save money. All this is due as much to lack of confidence between employer and employee as to the lack of intelligent views concerning the value of moral qualities in industrial classes. Workingmen as a rule know nothing of the financial world; their employers do. The language of the money market is an unknown tongue to the average mine worker, but his employer is conversant with it. Is there not here a sphere where the capitalists may render good service to their laborers? If simple lessons were given mine workers in money, its nature, its value, its use, etc., many of them would possibly be induced to save and try by experiment to verify the instruction imparted to them.

Thrift must be taught, and the sooner the lesson is inculcated, the better. It should be taught the child in the schools by instituting school savings banks. They are much needed. One of the most prevalent evils among our people is improvidence, and society will be strengthened if, by any system, the youths can be taught to economize. Children raised in homes where parents spend their spare cash in momentary gratification of appetite must be taught economic foresight. This is the reason why we have so many thriftless young people in our towns. If the public schools, by the simple method of the school savings banks, come to the aid of these children, they will be far more efficiently prepared for the duties of life than if they enter upon them equipped only with theoretic knowledge.

In the anthracite coal fields, outside the city of Scranton, this beneficent institution is not found. The system was adopted in the above city in the year 1897. It was first established in one of the schools and, during the year, five others adopted it. At the close of the year $950 had been received in deposits from the children. One of the rules governing the institution is that the children must in some way earn the money which is deposited. The teacher, each Monday morning, gives ten minutes time to receive deposits. These range from one cent up, and when a pupil's account reaches a dollar, he receives a bank-book from the bank where the money is deposited. At the close of the year the superintendent who introduced the system in the above schools said: "There is no tendency toward injurious rivalry among the pupils and no interruption of school work." At present, largely because of the strike of 1902, there are only two schools in the city continuing the system.

In our discussion of the boys in the anthracite coal fields we saw what is the custom among them on pay-night. All the money they get is "blown in" that evening. It is pure dissipation. One of the store-keepers who sold to the small boys of a mining town of 6,000 population said, that he took in as high as a thousand dollars a month when the mines were working regularly. The practice among girls is the very same. They take their pennies, nickels and dimes and spend them immediately in self-indulgence. A population that is thus from its youth accustomed to spending all upon the momentary gratification of appetite, will not learn the blessings of self-renunciation which is the foundation of strong manhood. Miners are indulgent to their children and the money spent on sweets, soft drinks, peanuts, and ice cream by the youths of the anthracite coal fields is over $75,000 annually. Here then is a promising field for teachers in public schools, for pastors and priests, to work in. Let the youths be taught that the practice of economizing fosters self-denial, strengthens character and produces a well-regulated mind. Thrift is based on forethought which is the one thing necessary to foster economic foresight; it makes the work of temperance comparatively easy and prudence becomes the dominating characteristic in life; it builds a barricade against vexations and anxieties which harass and perplex and secures comfort when declining years come; it drives away care and gives virtue the mastery over self-indulgence. These are blessings which are sadly needed in the life of our people and secular teachers and spiritual leaders cannot be engaged in better work than in inculcating these virtues in the rising generation. The school savings bank is one of the best means devised for this purpose, for charitable organizations, which in some cities do admirable work in this respect, have not commenced work of this kind in the coal fields. Returns from schools where the savings bank system is practiced, show that two fifths of the pupils become depositors. In the city of Scranton 60 per cent. of the pupils were depositors. The system should be introduced in all our schools.

The anthracite coal fields are also very poorly provided with means whereby the adult, who desires to save the few dollars which remain after his monthly bills are paid, can put his cash in safe keeping. We have many banks in our territory, but these are invariably located in towns and cities, far removed from mining villages where the majority of our people lives. It is nothing unusual to find strong iron rods fastened across the windows of dwellings in mining camps, which resemble the windows of prisons and banks, and they serve exactly the same purpose as bars and bolts do in banking establishments. The tenants are paid every two weeks and they stow away their spare money in trunks or drawers, and the iron bars are for the purpose of keeping out thieves who have often stolen money from these homes. Sclav bachelors generally keep their money in the house and instances are not rare when villains have stolen trunks, carried them to the brush, rifled them and pocketed the money they found. In the neighborhood of Shenandoah, a gang of thieves preyed upon the houses of mining patches for years and many were the victims of their raids. One mine foreman lost over $900 in one night. It is foolish to keep such an amount of money in the home, but we must remember that these men live far away from banking conveniences and work seven days in the week. The wives also are busy and cannot go to town. When the men come to town it is generally in the evening and then the banks are closed and the saloons are wide open.

Many of our people are thrifty and save a part of their wages. Many more could do so if facilities were given them safely to deposit their spare dollars and provide for possible contingencies. If such conveniences were instituted we are sure that much of the money now spent uselessly would be saved and sobriety would be greatly increased among the employees. The multiplicity of saloons is now one of the chief causes of thriftlessness among our people, and the antidote is the multiplicity of savings institutions.

Every one interested in the welfare of the working classes cannot but regret the failure of Mr. John Wanamaker, when

Postmaster-General of the United States, to introduce the Post-Office Savings Banks Institution. The banking interests of the country opposed the movement and showed clearly that pecuniary interest is stronger than patriotic sentiment. One thing the rich must learn, namely, that the tenure of their enjoyment of the wealth they possess depends upon the distribution of riches among all classes in the social hierarchy. Extreme poverty among the masses and superfluous riches in the hands of the few, will inevitably result in ruin. The rich owe it to themselves and to their country to give all possible encouragement and opportunities to the working classes to save, and the Post-Office Savings Bank is one of the best instruments for this purpose. These would afford the people a savings institution that would be open as long as the post-office is, generally from 7 a. m. to 7 p. m. The security given the workingmen would be equal to that of the Government, and the depositor would be able to get his money in any part of the United States, regardless of the place where he deposited it. It is not to America's honor that she and Germany are conspicuous examples among the civilized nations of the earth in not providing this means of saving for their working classes.

The need of such an institution in the anthracite coal fields is apparent from the accompanying map, which shows the location of towns in which banks are found, and also towns and villages where no opportunity is afforded the mine workers to save. From Forest City, in Susquehanna county, to Scranton, a distance of thirty miles and where 60,000 persons live, the only towns in which savings banks are instituted are the above-mentioned ones and Carbondale. Another bank has been opened this year in Olyphant. From Scranton to Pittston, a distance of ten miles, where 25,000 persons live, there is no bank, and from Pittston to Wilkesbarre, another ten miles, having a population of 31,000, there is no bank. On the Hazleton mountains, the only two towns having banks are Hazleton and Freeland, while around these are populous mining camps. In Schuylkill county, outside Pottsville, Tamaqua, Mahanoy City, Shenandoah and Minersville, there is no bank

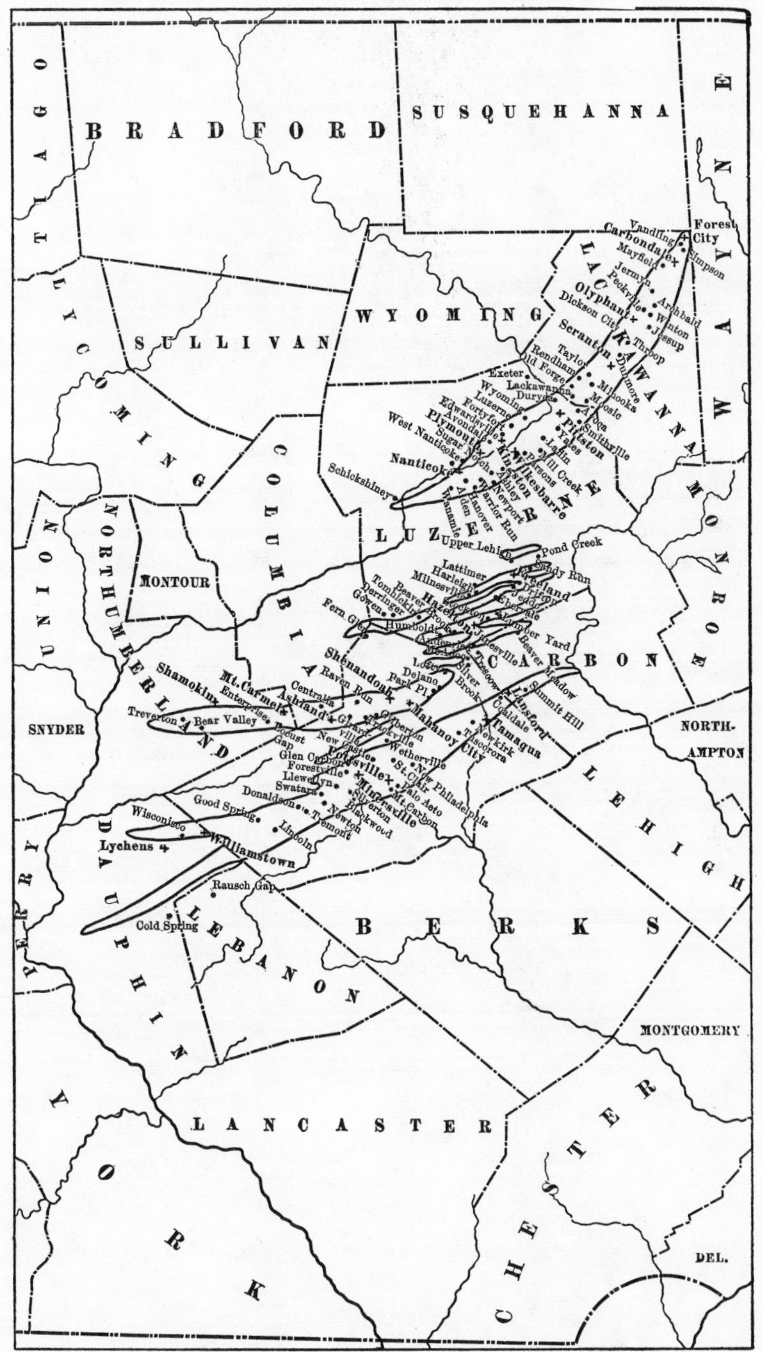

MAP OF COAL FIELDS SHOWING TOWNS WITH AND WITHOUT SAVINGS BANKS.
Towns marked (x) have Savings Banks in them.

in the mining districts. In all mining towns there are post-offices, and if in connection with these savings banks could be established, it would afford the man with a few dollars to spare, the opportunity of saving them. There are no post-office savings banks, however, and no likelihood that they will be soon established. Some individuals in mining towns conduct private banking. In the absence of an institution which would furnish reasonable safety to the savings of workingmen, could not our mining corporations devise a scheme whereby the employees could be aided in an effort to lay something by for old age?

The banks which are in the anthracite coal fields reflect the general financial condition of our territory. A banker, in his fifties, who had spent his days in the banking business, said: "I would prefer to start a bank in the anthracite coal fields to any other section. It's choice territory. There's lots of money circulating." The deposits in the banks show this, the major part of which represents the savings of mine employees.

There are in the anthracite coal fields 54 banks, 18 of which are in the cities of Scranton and Wilkesbarre. The total deposits in these banks, according to the reports issued January, 1902, were $50,164,728.56, which was 9.2 per cent. of the total deposits in the banks of the State and which approaches very nearly the percentage of population of the State in the anthracite coal fields, which is 9.8. The deposits in all the banks of the State amounted to $86.52 per capita of population; that of the banks in our territory to $79.62 per capita. In the cities of Scranton and Wilkesbarre the deposits amounted to $178.45 and $229.64 per capita respectively. Outside these two cities, there were 36 banks which had on deposit an amount equal to $66.28 per capita of population. Of course, all the deposits in our territory do not belong to the mine workers. Many bankers put their share at 50 per cent. of the whole. We will probably be nearer if we put it at 45 per cent. This would leave to each mine employee in the anthracite industry $152.85.

There is a great difference in the total deposits in banks in mining towns when computed in per capita sums of their population. The following table gives the per capita amount in the banks of anthracite mining towns.

Place.	Deposits per Capita.	Place.	Deposits per Capita.
Lykens	$ 91.45	Carbondale	$188.93
Williamstown	29.96	Pottsville	213.09
Mt. Carmel	46.13	Pittston	168.00
Mahanoy City	62.17	Hazleton	268.27
Nanticoke	24.29		
Plymouth	19.78	AGRICULTURAL BOROUGHS.	
Shamokin	73.87	Indiana	$237.67
Shenandoah	38.29	Mifflinburg	68.23
Forest City	5.78	Middleton	47.71
Minersville	55.93	White Haven	178.82
Lansford	16.73		
Freeland	89.84	BITUMINOUS BOROUGHS.	
Ashland	70.71	Du Bois	$ 94.85
Tamaqua	28.15	Scottdale	96.37
Kingston	66.90	Greensboro	366.32

We add for the sake of comparison four agricultural boroughs and three boroughs from the bituminous regions. Taking these generally, we find that agricultural boroughs compare very favorably with those in the coal fields. Wherever a town forms the county seat or has mixed industries the amount of deposits is larger. This is seen to be the case in both columns. Lykens, Shamokin, Kingston, Pottsville, Pittston, Carbondale and Hazleton show this clearly, as do also Greensboro and Indiana. Towns also in close proximity to cities show a small deposit, for many persons prefer to take their money to the city bank. Other banks in towns surrounded by many small mining camps have a high per capita deposit, for the thrifty from these surrounding villages bring their money there for safe keeping. These varying conditions make it almost impossible to get an accurate comparison of the money saved in mining towns as compared with agricultural or bituminous regions. All know, however, that mine employees do not practice thrift as do farmers. If they did, the banks in the anthracite coal fields would far excel in amount of deposits any others patronized by the working classes in the State. If a moiety only of the sum annually spent by mine workers on drink and tobacco were saved, the deposits in these banks would be doubled in ten years. As it is, our territory compares favorably with the State at large, and taking into consideration the number of banks in these coal fields, it compares favorably with agricultural communities. Another factor in this computation is the large number of aliens among us. Many of the Sclavs and practically all the Italians are sojourners only; they have come to make money, save it and return to their native country. These send their money regularly to the fatherland and longingly look forward to the day when they can return to the land of their birth. The Sclav and Italian form about 45 per cent. of the mine workers to-day, and the money they send to their friends across the sea would not fall much below half a million dollars annually.

Thrift is practiced by a large number of the Sclavs and Italians, but it has fallen into discredit among the Anglo-Saxons. The socialistic tendency of the time is fatal to the practice of this old virtue. Men are taught to look at consumption only and disregard production. All are anxious to get as much as possible of the good things of this life, and many spend their wages on glued furniture, gaudy apparel and vaudevilles. This wild passion, if persisted in, will only lead to one result. Those who have nothing are demanding and will demand more and more that which others have saved, and a fatal paralysis will fall upon the productive energy of the people. Our only way of escape is that the common sense of the working classes will lead them to see that production as well as consumption must be looked after, and that the industries of a country will only prosper in the fructifying soil of capital which is unconsumed production. John Graham Brooks says that the coöperatives of Belgium have come to the conclusion "that the members see the necessity of saving, borrowing, lending, even if in form they violate every theoretic principle of socialism." It is well that they have reached that conclusion, and sooner or later the workingmen of America will also come to the same conclusion. "Self-help, forethought, and frugality are the roots in personal character which nourish and sustain the trunk and branches in every vigorous commonwealth."

THE "BUILDING AND LOAN."

In our State the total assets of building and loan associations decreased in the year 1900 over a million and a half as compared with 1899. In the year 1901 they increased over three hundred thousand dollars. Of the total assets 7.6 per cent. were of associations in the anthracite coal fields, which was less than our percentage of the population of the State.

Building and loan associations began to flourish in our country as early as 1840. At that time the anthracite regions were little known and the tonnage sent to market from these coal fields was small. It was not until the early seventies that building and loan associations were organized in anthracite mining towns, and the citizens of the Southern coal fields were the first to introduce them. The following table shows the approximate date of the organizations of the associations now existing in these coal fields, classified according to counties.

Year.	Lackawanna.	Luzerne.	Schuylkill.	North-umberland.	Carbon.	Dauphin.	Columbia.
1870–1875			5	1		1	1
1875–1880			1	1			
1880–1885			4	2			
1885–1890	1	2	3	1			
1890–1895	13	6	3	3	2		
1895–1900	6		6	3			
	20	8	22	11	2	1	1

Of the three counties where anthracite mining forms the staple industry, Luzerne stands in striking contrast with the other two as to the number of building and loan associations. Five of the eight associations in this county are in Hazleton, while Wilkesbarre with its 51,000 population had only two. Scranton had 17 associations, while Ashland in Schuylkill county, with a population of only 6,500, had 4. Lackawanna county has entered this line of business very extensively in the last 10 years, while Schuylkill county shows a steady growth for the last 25 years. These associations have also flourished in the mining towns of Mt. Carmel and Shamokin, Northumberland county, where 11 of them exist. The four associations in Carbon, Columbia and Dauphin counties are in mining towns.

Of the share-holders in the 65 associations in the anthracite coal fields, the larger proportion is made up of mine employees. In the report of the Commissioner of Labor for 1893 on "Building and Loan Associations" we are told that 69.96 per cent. of the whole number of share-holders in local associations was practically working people, while 54.06 per cent. in the national associations belonged to that class. The commissioner says "these figures show conclusively that the building and loan associations of the country are being used by the class for which they were originally established." In a computation made of mine-working share-holders in purely mining towns, we found the percentage to be 54.6.

BUILDING AND LOAN ASSOCIATIONS.

County.	Number of Associations.	Total Assets.	Portion Credited to Mine Workers.	Percentage of Mine Workers' Share in County.	Percentage of Mine Workers in County.
Lackawanna	20	$3,191,231.03	$1,742,412.14	37.5	23.9
Luzerne	8	392,211.89	214,147.69	4.6	34.6
Schuylkill	22	3,111,595.98	1,698,931.40	36.5	23.0
Northumberland	11	1,557,628.03	850,464.90	18.3	9.6
Carbon	2	150,649.21	82,254.46	1.8	4.3
Columbia	1	44,321.03	24,199.28	0.6	1.5
Dauphin	1	55,449.98	30,275.68	0.7	1.5
Totals	65	8,503,087.15	4,642,685.55	100	98.4

The total assets of the 65 associations in the anthracite coal fields were $8,503,087.15. If the proportion above stated holds true of the territory under consideration, the mine employees are credited with $4,642,685.55. The above table gives the number of associations by counties, the total assets, the amount credited to share-holders among mine employees, the percentage held in the respective counties by mine workers and the percentage of mine employees in these counties.

From this table it is seen that the working classes of Lackawanna, Schuylkill and Northumberland counties have availed themselves of this means of building themselves homes to a greater extent than in other counties in our territory.

In the report of the Commissioner of Labor above referred to, 455 associations in our State had helped parties to acquire 19,091 homes and 425 associations had aided persons to erect 2,328 other buildings. In the year 1900 the 65 associations in the anthracite coal fields had aided to put up 583 houses. We have no means of knowing accurately how many houses the associations of these coal fields have aided to build, but an estimate made from the data given in the report of 1893 as well as the number aided by our local associations in 1900, places the number at about 4,000.

The average age of associations in our country is 6.3 years; that of those of our State is 8.1, while that of the associations in our territory is 10.1 years. During the last 25 years the building and loan associations in Schuylkill county have aided many to secure homes, and the towns where workingmen have the opportunity to put up houses have greatly increased in assessed valuation. The following eight towns may be taken as specimens of the increase in real estate in the places specified.

Place.	Year of Incorporation.	Assessed Value of Property at time of Incorporation.	Assessed Value of Property To-Day.	Percentage of Increase.	Average Per Year.
Shenandoah	1866	$131,144	$1,781,000	1,257	34.9
Ashland	1857	288,435	955,950	231	5.1
Girardville	1873	143,180	401,681	180	6.2
Minersville	1843	192,188	839,249	336	5.7
Mahanoy City	1864	138,996	2,084,535	1,400	36.8
Gilberton	1873	101,003	236,288	133	4.6
Shamokin	1866	162,862	1,154,865	670	18.6
Mt. Carmel	1864	40,394	681,132	1,586	41.7

The table shows a healthy growth and in each instance the increased wealth is directly or indirectly due to the mining industry. The "unearned increment" may account for a portion of the increase, but the vastly larger portion represents the earnings of men who have directly or indirectly made their money in the anthracite coal fields of Pennsylvania.

The effect of industrial crises in these coal fields is well reflected in the organization of these associations. In the early seventies the anthracite industry flourished in Schuylkill county, but from 1874 to 1882 ruinous strikes paralyzed business, and during this period only three associations were started. Scores of mine workers fell into arrears and after a brave struggle lost their property, while the financiers of the funds bought in the claims at nominal prices. One of these men said the associations made over 50 per cent. profit on these lapses. Industrial strife affected building and loan associations precisely as it did coal operators; the financially weak were crushed and the strong grew fat upon the wrecks. In 1888 another ruinous strike came upon Schuylkill county and from that year to 1891 only four associations were organized. Over 64 per cent. of the associations now in these coal fields have been organized during the last decade, half of which were instituted in Lackawanna county. In the year following the strike of 1887–1888 in the Lehigh and Schuylkill regions, not less than 48 persons were sold out by sheriff sale. This, however, represents only a fraction of the failures. Workingmen struggle bravely against an arrearage — which each month bears 2 per cent. interest — before they give up in despair. Many of them make new loans and hope to be better able to carry the load by extending it over a larger number of years. In the last strike of 1902 over 200 working people in the town of Mahanoy City could not meet their monthly dues and, although few of them have been foreclosed, the anxiety and worry which came to these families will never be told. The penalty for falling in arrears — 24 per cent. per year — frustrates every effort at reclaiming lost ground, but not until this usury crushes all hope is the struggle given up and the sheriff allowed to sell the home which has long been the care of the family and the object of the father's ambition. We have seen men, who, during years of normal activity in the mines, were industrious and sober and looked forward with joy to the day when the simple home would be theirs, become drunkards and reprobates when the object of their labor and the dream of their life was sold under the sheriff's hammer and they and their families evicted. A building and loan association is an aid to workmen to secure homes, but to the unfortunate ones who fall into arrears it becomes a deadly parasite which mercilessly consumes the accumulation of years.

The fact that so many associations in these coal fields have been so well managed by men who do not profess to be financiers is highly complimentary to the honesty and capacity of the organizers. They have afforded every industrious man the means of securing a home. The managers of the associations have worked for low salaries and conducted the business on a sound financial basis. The greatest hardship comes in when the industrious, for reasons over which they have no control, fail to meet their premiums and see the struggle of years defeated by the usury associations charge for arrears. In all the years of anthracite mining few operators have aided their employees to secure homes. Coal companies have sold lots to their men, but no scheme, based on the principle of "five per cent. and the fear of God," has been launched in aid of the mine workers. Every man who owns a house gives his employer a mortgage for good conduct and industrial efficiency, and operators could have done much to secure the peace and enhance the morality of these communities, if they had advanced a plan whereby their employees could have secured homes and be safely guarded against misfortunes which befall many in this risky industry.

Greater thrift and economy among mine employees would have also resulted in the erection of many more homes. C. C. Rose, Superintendent of the Delaware and Hudson Coal Co., said, that the 30 collieries in his charge lost half a day after each pay. The loss in wages to the employees in one year aggregated $322,628. All the coal companies throughout the coal fields make the same complaint, so that the loss of work because of the drink habit means an annual loss in wages to mine workers of $3,226,284. This sum, if spent in building homes, would, in less than 25 years, secure a sufficient number of houses, at $1,000 each, to shelter the 400,000 persons de-

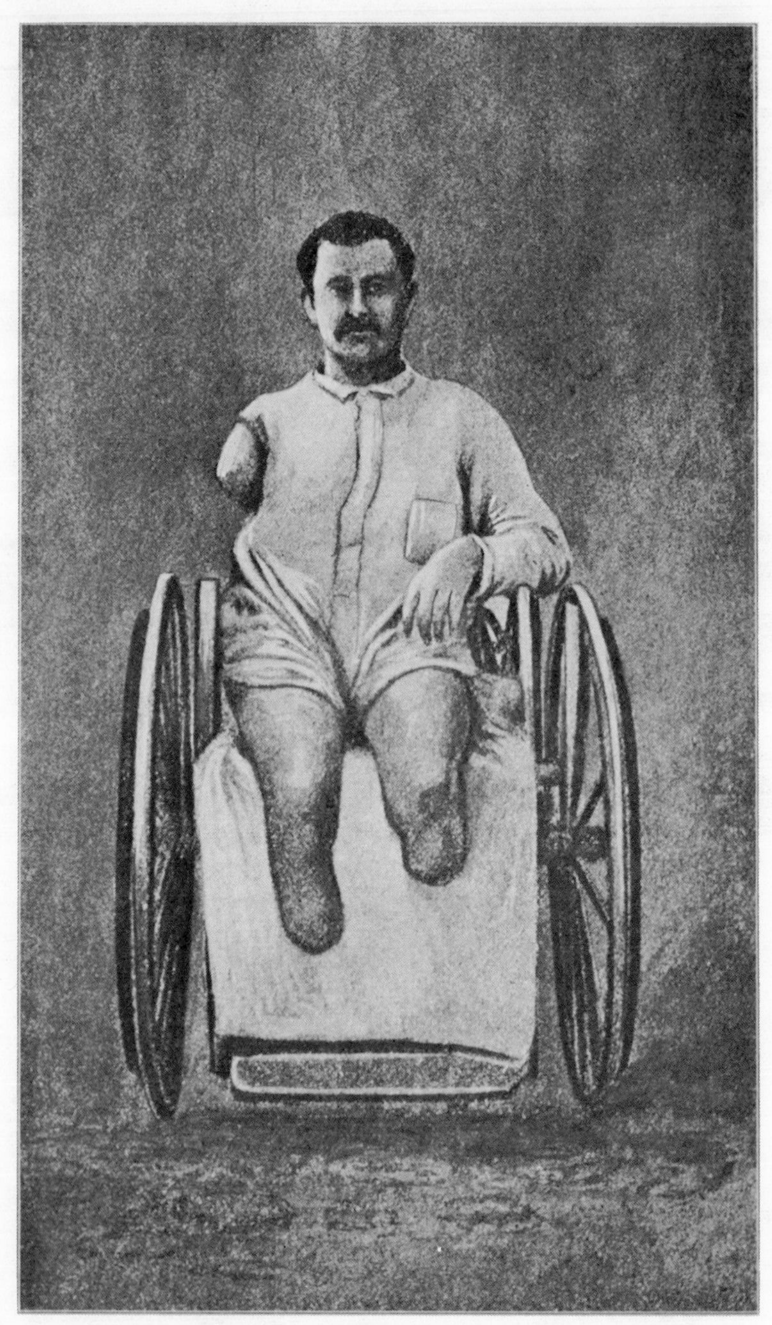

A Victim of a Non-fatal Accident.

pendent on the anthracite industry. Add to this the money spent in drink and it can be done in less than 8 years.

Organizations for Sick Benefits.

One of the signs of the economic well being of the working classes of America is the multiplicity of fraternal organizations among them. There is a plethora of orders of all sorts in anthracite mining towns. In the town of Olyphant, there were, in 1902, 31 societies of all sorts, or one for every 250 persons of the population. In Minersville there were 17, or one to every 280 persons. In Shenandoah, there were 60, or one for every 330 persons. In Hazleton, there were 52, or one to every 274 persons. In Nanticoke, there were 35, or one to 340 inhabitants. It is so throughout our mining communities. Organizations of all kinds for both male and female are found, which involve annually business transactions amounting to hundreds of thousands of dollars, and the business is virtually managed by the workingmen themselves.

There are many fraternal orders in our towns for purely social purposes, but these are supported by the business and professional classes. Orders which are supported by our working classes are distinctly for purposes of insurance against sickness and death. Men who earn their daily bread by arduous toil have no money to spend in orders of a purely social character. They fear economic loss through sickness and are anxious to provide some means against it. They are solicitous for a decent burial and regularly meet their dues that the funeral benefits may be paid. Beyond these two contingencies, the average workingman does not concern himself. An insurance man of wide experience in our communities said that 75 per cent. of the working classes carries some kind of insurance, of which 25 per cent. carries policies against sickness, and 50 per cent. carries policies of from $100 to $300 against death. Two thirds of these classes have no other provision against sickness and death. They live from hand to mouth and are only solicitous for money to meet their dues. At Christmas time and the Fourth of July, the cash for premiums is scarce, for they live beyond their means and for the following month or two they are reduced to bare necessities. They cannot regularly manage their income so as to secure medium comfort each day in the month. They move in rushes and it is either a feast or a famine. Their economic foresight sees two possible contingencies — sickness and death, and against these they provide.

There is a marked difference between the Roman Catholics and the Protestants in their relation to fraternal orders and the ways in which the ceremonies are conducted. All fraternal orders have a religious basis, but those among Roman Catholics are invariably connected with the church and the priest is the spiritual director. Hence in burial ceremonies the authority of the church is not interfered with and the priest alone administers the rites. Those among Protestants recognize the religious element and claim that their order is founded upon biblical principles, but they have their own chaplain to conduct the burial rites. Hence an inevitable competition if not conflict comes in, and the Protestant clergyman is generally put in an embarrassing position between the chaplains of fraternal orders who insist upon performing the sacred rites of burial. All the fraternal orders among Protestants have an avowed creed which is nominally theocratic. Many of them avoid explicit declarations as to the person of Christ and yet express the hope of resurrection and immortality which is distinctly Christian teaching. They have no avowed declaration of faith and apply epithets to God which ill harmonize with the best expression of religious faith of the twentieth century, and yet they appropriate biblical passages in their ritual, copy the prayers of the church and use Christian hymns and music. These sacred implements are often found in unworthy hands in the Church, but the promiscuous use of them by the average chaplain among fraternal orders in our mining towns destroys their meaning and breeds contempt for sacred usages and customs. No order, which exists for beneficiary purpose in sickness and death, can discard the religious impulse. In the presence of death the human heart bows before the mystery of the ages, and the Christian expressions of hope beyond the grave which are contained in the Church's prayers, hymns, songs and Scripture, are the highest ever attained by man. If these are indiscriminately used by illiterate and unworthy men, a spirit of contempt for religious rites and ceremonies is fostered, which bodes no good for society. We have seen worthy laymen in charge of burial services and their earnestness and devotion inspire reverence and humility, but we have also seen men who resembled Nadab and Abihu in their handling of sacred things. It is difficult to see how to remedy this sacrilegious usage if the clergy persists in keeping aloof from these organizations. There is no reason why pastors cannot join fraternal orders and assume the duties which the members feel they ought to discharge at the burial of a deceased brother. The supplanting of the clergy in the ministrations for the dead injures none as it does religious organizations, and the remedy is largely in the hands of the Protestant pastors themselves.

Some of these orders also come in conflict with the Church in other spheres than the burial of the dead. On lodge night the Bible is used, prayer is offered and a hymn is sung, and that is all the religion many of the members think they need, except as the organization attends divine service once a year. Whatever ethical and religious culture these orders afford their members society is the better for it. But whoever substitutes the order for the Church knows little of the true place of the latter in society. Remunerative altruism, a vague and lax belief in God, an undeveloped religious sentiment, an artificial and crude ritual, and an undefined moral obligation restricted by no norm of conduct, are poor substitutes for the Church. Natures which resent the censor of the church for wrong-doing, and abdicate the religion which seeks "to save," find what they imagine satisfies their religious nature in religious fraternal orders. But society can ill afford to substitute undefined religious conceptions for the definite teachings of the Church and the good fellowship of a secular order for the norm of Christian conduct maintained by the Christian conscience.

Among loyal Catholics there is no conflict between the fraternal orders and the Church. Spiritual direction rests wholly in the hands of the priest, while the laity is left in charge of financial affairs. The Sclavs have many orders which are invariably called after the name of some hero or saint. Most of the societies are beneficiary and insure the member against sickness and death. Among the Irish and Sclav Catholics beneficiary societies are also instituted for women, while the Daughters of Rebecca, the Daughters of Pocahontas, and various other auxiliaries furnish the same to the Protestant women. Among Catholics there are fewer orders than among Protestants, and for this reason the local organization among the former is numerically stronger than among the latter. Individualism runs to extremes among Protestants in matters of fraternal orders as it does in religious organizations, which increases the cost of maintenance and imposes heavier burdens upon the members.

The finances of these orders are managed by the men themselves, and when we consider the vast sums handled annually by wage earners, the fact that instances of embezzlement are very rare is highly complimentary to the integrity of these men. It is impossible to state accurately the amount of money annually contributed to these orders in our area. In the territory of Olyphant (7,800 population) we estimated that the lodges received monthly about $1,886 in dues, which ranged from 25 to 60 cents. In addition to this many of the orders assess each member 50 cents or $1 when one of the order dies. In the towns located in the anthracite coal fields the fraternal orders flourish and receive in dues annually over a million and a quarter dollars. The benefits they pay vary from $4 to $6 a week for not more than from three to six months. If at the expiration of the first period, the recipient is still sick, the allowance is reduced one half and again runs for another three or six months. Beyond this the lodge does not provide so that if the patient is in need he becomes an object of charity and is aided either by the poor board or the generosity of neighbors. In case of death lodges pay from $50 to $125, and half rates to members when their wives die. Sclavs, who carry life insurance, get half the amount they carry in case the wife dies, but the society reduces this one third if wife number two dies, and one sixth if wife number three passes away, while no provision is made for number four in case of death. Sclavs who carry an insurance of from $300 to $600 are treated with considerable respect in case of death, but a poor Hun, brought home from the mines on the stretcher, may find a bolted door. "Dead man no good," they say.

This effort of the working classes to raise a barrier against pauperism and bring the various members of their society into relations of mutual helpfulness and friendship is praiseworthy. All adults among mine workers belong to some lodge which insures against sickness and death. But one cannot help observing that the multiplicity of lodges maintained by the working classes is a waste of energy and money. The object they aim at would undoubtedly be better served if greater concentration were practiced and the societies covered a wider range of territory than they now do. A more efficient system would thus be instituted and the coöperation of workingmen for mutual protection would be established on a firmer financial basis. In this way individual or local risks would be distributed over a larger number of men and communities subjected to a high death rate could be more efficiently aided. Men of wisdom and large experience in financial affairs could also render material aid to these societies of working men if they, by genuine democratic sentiment, became members and devoted a part of their time to guiding the affairs of the societies. It is true that they may be met by suspicion and find the workingmen shy and sensitive, but persistent and patient devotion to their welfare would in time remove the barrier, and an unselfish and earnest man can greatly add to the happiness of his fellowmen by such services.

These orders also furnish an admirable sphere for the exercise of the disposition of sociability. They are the expression of the social action of the members in distributing individual risks. The workingmen find pleasure in their lodges because the management of affairs is in their hands. Their meetings afford occasion for discussions and in the long run result in refinement of manners and in increased intellectual power. The results attained by employees in the management of their affairs may not be the highest, but they gain experience thereby and acquire business tact and an insight into the nature of the economic world which are of greater social value than financial considerations. Independence, self-reliance and foresight are the qualities which lift men in the scale of social rank, and in no sphere are workingmen taught them better than in the management of the business of these fraternal organizations.

Insurance Against Mine Accidents.

John Graham Brooks says in his interesting work, "Social Unrest": "In the anthracite coal fields one would like to begin reform by applying this systemized insurance to that frightful list of stricken laborers that are now thrown back upon themselves or their families with recompense so uncertain and niggardly as to shock the most primitive sense of social justice." All familiar with the risks in the anthracite industry and with the varied and unscientific systems of insurance which prevail, are in perfect accord with this sentiment of the eminent economist. In the year 1901 the fatalities among railroad employees was 2.5 per 1,000 and 3.5 per 1,000 among anthracite mine workers. Among switchmen, flagmen, and watchmen the killed was 5.3 per 1,000 employees; among our miners and laborers it was 5.5 per 1,000. The fatalities among miners in 1901 was 6.2 per 1,000. In the Southern coal fields the fatalities are more numerous than in the Northern. In the year 1901, the per thousand killed among the miners of the Philadelphia and Reading was 7.6, while in three of the collieries in that year the fatalities among the inside employees were 9.3, 10.8 and 11.7 per thousand respectively. For the years 1870–1901, the number of the killed in the anthracite industry was 10,318, while the number of those non-fatally injured during the same period was 27,311. But notwithstanding this great loss of life and limb no systemized prin-

TABLE OF NON-FATAL ACCIDENTS.

	1900.	1901.		1900.	1901.
Legs fractured	268	314	Ribs fractured	35	36
Arms fractured	80	97	Hips injured	36	29
Burned	201	245	Pelvis fractured	2	0
Bruised	220	219	Knee fractured	6	10
Skull fractured	15	19	Internal injuries	4	0
Face cut	21	72	Collar bone fractured	10	12
Teeth knocked out	4	1	Ankle dislocated	12	23
Eyes injured	11	6	Toes fractured	13	0
Jaw-bone fractured	7	6	Foot injured	35	46
Shoulder dislocated	6	9	Wrist broken	3	2
Shoulder fractured	9	6	Spine injured	5	1
Back injured	59	62	Thigh fractured	0	7
Hands crushed	19	41	Cheek-bone fractured	0	1
Fingers cut off	20	0			
Abdomen injured	6	0	Total	1,107	1,264

ciple of insurance has been adopted, and virtually this risky business has escaped the expense of accidents which justice demands should be counted as part of the costs of the industry. Many of our coal operators can say with Mr. Schwab: "Where an accident results fatally, the family is always taken care of financially. If there are children, provision is made for their education. If we cannot provide means by which the wife can take care of herself, we allow her a pension, or house to live in, or something of that description." Other operators care nothing for the fatally or non-fatally injured in their mines and make no inquiry as to the condition of the family of the unfortunate. All the countries of Europe where industries flourish, have taken steps to establish a system of industrial insurance which is just to the employees. Such a system is sadly needed in the anthracite industry and the cost should be placed on the coal sent to market, in the mining of which so many of our fellowmen are killed or maimed. When justice will be established in our industrial life, every risky business will be compelled to pay insurance proportionate to the peril to life and limb incident to it.

Our mine inspectors invariably say in their reports that over 50 per cent. of the fatal accidents in anthracite collieries is due to the carelessness of the men themselves, but evidently they do not allow for risks which are incident to the employment. In the tabulation of causes given by the Imperial Bureau of Germany in 1887, 43 per cent. is put down to the risks incident to the business in which they occurred, while only 25 per cent. was due to the fault of the victim. We doubt not but a similar conclusion would be reached by an impartial investigator into the accidents of these anthracite collieries. But whatever be the causes of these frightful accidents the victims should be aided and not thrown upon the charity of society, friends or relatives.

There are four systems of insurance among the mine employees of the anthracite coal fields. One form is that instituted by the employees themselves and wholly maintained and managed by them. The one in vogue at the Lackawanna colliery in the Northern coal fields is a good example of such a fund. The second form is that instituted, maintained and managed by the company. The one in vogue at the collieries of the Cross Creek Coal Company is a good example. The third form is that which is maintained wholly by the employees but managed by the operators. The fund in vogue among the employees of the Philadelphia and Reading is an example of this. The fourth kind is that maintained and managed by the coöperation of employers and employees. The fund of the Delaware and Hudson is a good example of this method. Besides the above forms of insurance a fifth may be added which is known as the "keg fund." The miners preserve the old powder kegs and get ten cents each for them; this money together with monthly dues is placed in a fund for accidents. Most of the shafts of the Delaware, Lackawanna and Western afford good examples of this form of insurance. Of all the anthracite mine employees about 60 per cent. would be in one or the other of the above forms of insurance, and the other 40 per cent. has nothing but the charity of operators or fellow employees to depend upon in case of need.

The forms of payment into these funds vary. In the Lackawanna colliery above mentioned, the adult employees pay 25 cents a month and boys under 16 years half that. When a member is killed a 25-cent assessment is levied per member.

In case of accident, $5 a week is paid for six months — if the injured is incapacitated that long — and $75 is paid in case of death for funeral expenses. Boys under 16 get half benefits. The annual expense connected with the fund is about $115, and over $1,000 annually is disbursed in benefits. The Philadelphia and Reading divides its employees into four classes and collects from each according to its scale of wages. In case of accident members are paid weekly sums proportionate to their contributions while provision is made for widows and orphans. Under the Delaware and Hudson the fund is maintained by an equal contribution from the employees and the company. Whenever the fund is low, the employees give their earnings of one day and the operators give an equal sum. In case of accident all persons earning over $1.20 a day are paid a dollar a day (except Sunday) for three months if the injured is incapacitated that long. Those earning less than $1.20 a day are paid 50 cents a day for not over three months. Contributions are voluntary. A committee of five at each colliery, comprised of three contributing members, the inside and outside foremen, looks after the injured. In fatal accidents $50 is paid for funeral expenses and in case a widow is left, $3 a week is given her and $1 a week to each child under 12 years of age, for one year. Of the 12,500 employees of the company 5,581 belonged to the association in January, 1903.

The first fund established in these coal fields was in 1869 by the Wilkesbarre Coal Company after the Avondale disaster. It was operated after the manner above described in the Delaware and Hudson fund. Charles Parish established a similar fund in the Lehigh and Wilkesbarre Coal Company in 1873. The fund lasted one year and was abandoned by the company for "the men abused it." In 1875 the Lehigh Valley Company instituted a fund on the same plan as that of the Delaware and Hudson. In the conflict of 1887–1888 the employees got suspicious and the employers indifferent, so that the fund was dropped in the collieries of the company in the Middle and Southern coal fields. It was kept going in its collieries in the Northern coal fields for a few years longer, but it is wholly abandoned to-day. The Philadelphia and Reading established its fund in 1877, and the Delaware and Hudson in 1887. On the Hazleton mountain the custom generally is for the operator to contribute $50 in case of death and the workmen 50 cents each. No insurance is provided for non-fatal accidents in this territory, but employees dealing in company stores are "carried" while incapacitated. The Coxe Brothers, as before mentioned, give $5 a week to the injured. This company is liberal in the matter of insurance, and one of its employees said: "Well may they for we work for them on the calico basis." * Lower wages are paid day employees under this company than under any other in the coal fields.

There are many instances of employees refusing a scheme of insurance proposed by operators. The Pennsylvania Railroad Company proposed a system of insurance to its employees in Williamstown and Lykens, modeled after its scheme for its railroad employees, but the miners refused it. The Lytle Coal Company proposed one modeled after the manner of the Delaware and Hudson, but the men were suspicious of the good intentions of the employers and refused it. The employees of the Philadelphia and Reading refused the scheme which was proposed them in 1873, and it was only by the determination of Franklin B. Gowen, who knew better than the men themselves what they needed, that the present system was established. It began by a contribution by the company in 1875 of $20,000. It has worked well for over 25 years, but at present many employees complain of the tyranny of the corporation which, they say, annually collects thousands more than it pays out. They think not, or possibly they are ignorant of, the deficit of former years which is gradually being wiped out by the surplus annually collected.

The Lehigh Coal and Navigation Company's fund has upon it the stamp of the late Abram S. Hewitt, who, in 1877, in-

*The "calico basis" refers to an agreement made in the seventies between the employees and E. B. Coxe. Mr. Coxe informed the men that he could not compete with other operators unless they consented to a reduction. This reduced their wages so low that the wives of the miners could get nothing save calico to wear, and the agreement was called the "calico basis."

sisted upon the company, of which he was a director, doing something for its employees, many of whom lived like "hogs and dogs." In 1883 the company donated $20,000 to found a fund and then levied one half of one per cent. on the wages of inside labor for the maintenance of the fund, while the company, as its portion, contributed a fixed sum for each ton of coal mined. The sum of $30 is paid for funeral expenses, half pay is given the widow or family of the killed for 18 months, and half pay is given the injured for not more than six months.

The following table compiled from the annual reports of the Delaware and Hudson gives the number of injured, killed, etc., for the years specified, in the collieries of this company.

Year.	Number of Injured.	Average No. of Days Lost.	No. of Single Men Killed.	No. of Married Men Killed.	No. of Orphans Left Under 12 Years.
1887	101	12.81	3	2	6
1888	259	16.62	2	6	16
1889	359	16.81	6	3	8
1890	329	15.81	1	2	2
1891	506	15.72	4	5	8
1892	748	14.80	0	4	4
1893	989	14.48	5	4	15
1894	867	13.92	3	5	9
1895	1,171	14.13	5	8	12
1896	1,405	14.03	6	9	23
1897	1,132	14.94	11	10	28
1898	637	17.51	3	10	13
1899	686	18.48	9	9	12
1900	606	18.97	2	14	20
1901	673	19.02	9	8	15

The table shows clearly that non-fatal accidents are of a more serious character in recent years than in previous ones. This is due partly to the installation of heavier machinery and partly to the larger amount of powder used in the harder and smaller veins.

During the existence of this fund from 1887 to the close of 1901, the company disbursed the sum of $197,466.53, and collected $206,454.58, so that there was left in the treasury on January 1, 1902, $8,988.05. The company annually publishes a statement, giving all necessary details as to collections and disbursements. This is commendable and invites the confidence of the employees among whom the statements are distributed.

The Philadelphia and Reading Coal and Iron Company has the most extensive system of insurance against accidents of all the operators in the anthracite coal fields. From 1877 to 1900 it was practically compulsory, but during the recent industrial friction there has been an assertion of individual rights on the part of some employees, which results in overt criticism of the fund. This has had a discouraging effect upon some of the officials, who have seriously considered the advisability of discontinuing it. No more serious mistake could be committed for both employers and employees than to discontinue the fund before some other means of relief is provided against accidents.

The fund was established, as above stated, in 1875 by an endowment of $20,000 by the company which was invested in first mortgage bonds in the Mammoth Coal and Iron Company and the Preston Coal and Improvement Company, bearing 4 per cent. interest annually. The employees were then classified into four classes and the sums of 30 cents, 20 cents, 10 cents, and 5 cents a month kept at the office of the company for the maintenance of the fund. The benefits were $5 a week for not more than six months to those of the two first classes; $2 a week to those assessed 10 cents, and $1 to those assessed 5 cents; $30 was paid for funeral expenses; $7 a week was paid to the family of men killed in the first and second class; $2.80 to those of the third class and $1.40 to those of the fourth class, for one year. This scheme worked for 12 years, but the dues were not enough to cover the disbursements, so that at the close of 1888 there was a deficit of $131,275. This was the year of the sympathetic strike of the Philadelphia and Reading men, and possibly the ingratitude of the employees had something to do with the resolution of the company to make the fund self-sustaining. In 1889 the dues were advanced to 50 cents, 40 cents, 30 cents and 15 cents. The benefits were also changed, and $5 a week was now paid to men in the first, second and third classes and $2 to those in the fourth class; $30 was paid for funeral benefits; $7 a week to the family of those killed in the first, second and third classes, and $2.80 to that of those in the fourth class for one year. This scheme from the first year left a balance which could be applied to the deficit, so that from January 1, 1889, to December 31, 1902, the sum of $127,053.43 was paid on the debt of the old fund, leaving still the sum of $4,710.07 to be paid. During the existence of this fund the company has paid to the 73,821* injured employees $1,344,769.32, and to the families of the 3,019 killed, $577,724.74. The maximum number of contributors to the fund was in the year 1897, when it reached the total of 27,682 employees. In 1900 the number was 25,541, and in 1901, 23,254. The presence of the Miners' Union in the coal fields as well as the irritation caused by the strikes of 1900 and 1902, has resulted in strained relations between employer and employee, and the fund, which has accomplished great good in the last quarter of a century, is subjected to criticism. The employees in their conventions have openly charged the company with collecting more than it disbursed, and condemned it as unjust "in keeping thousands of the employees' wages every year." The following statement, published in a report of one of their conventions, shows to what extreme suspicion leads men: "It is alleged that the company [P. & R.] collects about $120,000 per annum from its workmen in the Ninth district, and that less than 25 per cent. of

*This number of injured among the employees of the Philadelphia and Reading Coal and Iron Company ill harmonizes with the total injured among all employees of anthracite collieries as given on page 264. The discrepancy is explained by the fact that the reports of mine inspectors, from which we have taken the number injured for 31 years (1870–1901), do not record accidents as accurately as do the ledgers of accident funds. During the decade, 1891–1901, an average of 3,684 persons among the employees of the Reading Coal and Iron Company, annually drew claims because of accidents. This rate gives one out of every seven, or 147 persons per 1,000 employees non-fatally injured each year. The accidental fund of the Delaware and Hudson shows that in the decade 1891–1901, an annual average of 891 persons was paid benefits, which was one out of every seven members of the fund, or 148 persons injured per 1,000 employees. Among railroad employees 1 is injured to every 29, and among trainmen 1 to every 10 employed.

this sum is paid out in benefits." The company could allay much of this suspicion if it annually published a statement, such as is issued by the Delaware and Hudson Co., whereby the employees may be accurately informed as to the condition of the fund. However, all such transactions finally rest upon confidence between employer and employee, and, somehow, the presence of the union has largely destroyed this. One of the managers explained it by saying: "These men repose more confidence in unscrupulous leaders than in responsible business men." Blind ignorance, suspicion and prejudice characterize both parties. These disintegrating forces have too prominent a place in the ranks of both capital and labor, but we hope they will not prove fatal to the most extensive system of insurance in the coal fields.

The following table, compiled from the reports of the Philadelphia and Reading Co., shows the number injured, killed, etc., from 1887 to 1902:

Year.	Number Injured.	Average No. of Days Lost.	No. of Killed.	No. Killed per 1,000 Contributors.
1887	2,900	19.08	95	5.8
1888	2,901	18.06	109	5.5
1889	3,089	20.16	111	5.6
1890	3,520	19.26	126	5.5
1891	4,274	18.78	153	6.7
1892	3,838	20.82	152	6.2
1893	3,696	22.80	139	5.7
1894	3,904	23.40	165	6.3
1895	4,189	23.52	164	6.2
1896	3,925	24.42	169	6.1
1897	1,951	27.78	141	5.1
1898	3,662	25.92	174	6.4
1899	3,986	25.43	149	5.9
1900	4,181	23.83	149	5.9
1901	3,840	24.56	158	6.1
1902	3,511	25.40	168	7.2
6 Months.	571	22.89	95	2.8

This table shows the same as did that of the Delaware and Hudson, that the character of the non-fatal injuries is more serious in the last decade than in previous years.

Of all needs in the anthracite coal fields there is none greater than that of a system of insurance based upon scientific principles and including all the employees of the industry. Nearly half the employees have no provision for either the incapacitated through accident or for the maintenance of widows and orphans when death befalls those who provide for them in this hazardous calling. Many operators display generosity worthy of emulation; others manifest criminal indifference to the sufferings of employees and their families because of accidents. There is no reason why any of these men, in case of accident, should be left to the uncertainty of the generous impulse of operators. Industrial friction is fatal to generous sentiment. One of the operators on the Hazleton mountain annulled a beneficent system of funeral benefits which had been in existence for 20 years because, as he claimed, "the union violated its contract," when in fact the men only acted from motives of sympathy toward a fellowman killed in the mines. To leave these men to the mercy of over-bearing operators in case of injury and death is unworthy of the civilization of the century in which we live. The employees deserve better treatment. A system of insurance modeled after that of Germany or France should be drafted and enforced, the basic principle of which is that the business should pay risks commensurate to the perils surrounding the men digging coal. It is unjust to impose the burden wholly upon the employees; the employers also should bear a portion, if not the greater part of the burden. A general system incorporating the scheme practiced by the Lehigh Navigation and Coal Company would probably be the best possible one, in order to encourage care and vigilance on the part of employees and also throw upon every ton of coal mined a part of the burden due to accidents which is now borne almost wholly by the workingmen themselves. It has long been an established principle that production must provide for the depreciation of machinery; it is time it should also provide for the depreciation of man.

CHAPTER X.

OUR CRIMINALS.

1. Crime in the Anthracite Counties. 2. The Dockets of the Justices of the Peace. 3. Crime Among Farmers and Miners—A Comparison. 4. The Boys and Girls Who go Wrong.

Crime in the Anthracite Counties.

In 1900, 695 persons were admitted to the penitentiaries of our State ; 68.63 per cent. was committed for the first time and 31.37 per cent. recommitted from the second to the fourteenth time. The youngest of these convicts was only 16 years of age ; 11 per cent. between 16 and 20 years ; 42.86 per cent. between 20 and 30 years, and 46 per cent. over 30 years of age, while the average age of the group was only 31.9 years. Illiteracy was not high among them, for 87 per cent. could read and write ; 82 per cent. was native born and 18 per cent. foreign born ; but 85 per cent. had no trade. At the age of 16 years, 66.9 per cent. had both parents living, 25 per cent. had one parent living, and 8.1 per cent. had lost both parents. Only 16 per cent. was total abstainers and 43.6 per cent. was intemperate ; 63 per cent. was single or widowed at the time the crime was committed. Of the prisoners in county jails, about 2 per cent. was 16 years of age ; 59 per cent. between 20 and 30 years, and 39 per cent. over 30 years of age. Of these 82 per cent. was convicted for the first time, and the remainder recommitted from the second to the tenth time. Illiteracy is not great here again, for 79 per cent. could read and write ; 82 per cent. had no trade ; 14 per cent. totally abstained from the use of alcoholic drink and 18 per cent. was intemperate ; 62 per cent. was single at the time the crime was committed. One fact is evident from these data, namely, that educational advantages do not appear to have a deterrent influence upon persons predisposed to crime, for those who had had the advantages of a public school education formed by far the majority of criminals.

In the three counties, Lackawanna, Luzerne and Schuylkill, where anthracite mining forms the staple industry, the population increased during the last decade 25.4 per cent., and the number of convictions for crime during the same period increased 34.15 per cent. From 1890 to 1900 in our State the number of convictions kept pace with the percentage increase of population, but did not exceed it. The following table gives the percentage increase of convictions and of population in each of the above counties in the last decade :

	Increase of Convictions.	Increase of Population.
Lackawanna.	43.96 %	36.4 %
Luzerne.	40.35	27.8
Schuylkill.	20.87	12.2
General.	34.15	25.4

The figures show that the number of convictions increased most rapidly in Luzerne county. The number of convictions to the 100,000 population is given in the following table :

	1890.	1900.
Lackawanna.	81.63	106.79
Luzerne.	33.79	44.33
Schuylkill.	105.73	119.12
General.	69.75	84.47

In all the State it was 66.48 and 66.37 per 100,000 population for the above years, so that the number of convictions kept pace with population in the Commonwealth, but increased in the anthracite coal fields.

The number of persons charged with crime in the three counties in 1890 was 507.58 per 100,000, and in 1900 it had increased to 891.83. In the State the number of persons charged with crime in the years compared was about the same. Of the bills presented to the grand juries of the three counties in 1890, 57.7 per cent. was returned as true bills, and in 1900, 35.1 per cent., but in the State at large it was 76.5 per cent.

and 70.8 per cent. respectively. Of the persons tried in our counties 23.7 per cent. was convicted in 1890, and 25.3 per cent. in 1900, while in the State about the same percentages prevailed.

In 1890, in the three counties there were 89.45 persons per 100,000 population in county jails, and in 1900 there were 119.57. In all the State the figures were 109.64 per 100,000 population in 1890, and 119.87 in 1900, so that the proportion of inmates in county jails in our territory was about the same as in the county jails of the State at large. Crimes are generally classified as against person or property; the following figures give the percentage in the State and in the three counties for the years specified.

	Year.	Crimes Against Person.	Crimes Against Property.
In State......................	1890	43.99%	56.01%
" "	1900	45.37	54.63
In Anthracite counties.	1890	54.43	45.57
" " "	1900	50.09	49.91

The table shows that crimes against person are more prevalent in our territory than in the State at large. In the penitentiaries of our State 70.6 per cent. of the crimes committed by the inmates was against property and 29.4 per cent. against person.

During the decade a marked increase was observed in the following crimes: assault and battery, carrying concealed weapons, disorderly conduct, desertion, homicide, receiving stolen goods, resisting officer and seduction; while larceny, malicious mischief and fornication remained about the same. A preceptible falling off is observed in violating the liquor law and false pretense, due possibly to a less rigid enforcement of the law.

From the study of the nature of crimes committed in these coal fields, we find that the number against person is very large, which is due to passionate outbursts and lack of self-restraint. It is the characteristic of savages to be swayed by passionate outbursts of feeling uncontrolled by reason, and unfortunately the same is true of a large number of our people. It is also worthy of note that of the total sum of recognizance forfeited in the State in 1890, 68.15 per cent. was in the above mentioned counties, and in 1900, 66.50 per cent. The figures also bring forth one other characteristic which will become more clear in the progress of this chapter, namely, the small percentage of true bills which grand juries in the three counties return of all the bills presented to them. In the year 1890, grand juries of the State in general found 76.5 per cent. of the bills true ones, and the grand juries of these coal fields 57 per cent., but in the year 1900 the former returned to court 70.8 per cent. while the latter only 35 per cent. In 1901, 71.9 per cent. was returned by grand juries of the State and 41.7 per cent. in the coal fields. It is a general complaint in our courts that the justices of the peace return trivial cases which ought to have been finally disposed of in the local court. When these cases come before our grand juries they are ignored, but the ignorance and want of judgment of local magistrates increase the costs to the counties* and become a source of great annoyance, vexation and expense to innocent parties who are the victims of malice or envy. Lackawanna county is the chief transgressor in this regard. From it comes 50 per cent. of the persons charged with crime in the three counties, although it contains only 31 per cent. of the population of that area. Of all the persons charged with crime in 1900 in the counties specified, only 37.4 per cent. was brought to trial, while in the State at large there was 80.5 per cent. tried. Of the inmates in Lackawanna county jail in 1900, 43 per cent. was there by the summary conviction of the justices of the peace. The Secretary of the Board of Charities speaks of this evil: "many are hastily convicted here (Lackawanna county) by magistrates for trivial offences. This at all times is bad, but when done in the case of children, is criminal." It is criminal in 90 per cent. of these cases and is

*Judge C. R. Savidge, of Northumberland Co., in the October (1903) county court was so chagrined over the trivial cases brought before him, that he advocated the abolition of justices of the peace and the appointment of police magistrates. Trivial cases which should be disposed of in the local court are returned and the cost often amounts to from $50 to $100 to the county.

due to the callous indifference or gross ignorance of justices of the peace as to the effect of imprisonment on men.

If we turn to the financial side of the question we find that it cost the people of the three counties in 1890, 12.33 cents per capita to maintain the prisoners in the county jails, and in 1900 it was reduced to 10.10. In the decade, however, the percentage of expenses appropriated for salaries of officers increased nearly 27.8 per cent. In 1890 it was 29.27 per cent. of the total expenditure and in 1900 it was 37.44 per cent. This shows that our territory is in perfect accord with the State in increased appropriations for the efficient officers the politicians choose. The three counties differ as to the percentage appropropriated to salaries. Lackawanna appropriates 35.1 per cent. of the total cost, Luzerne 44.7 per cent. and Schuylkill 33.7 per cent.

In 1890 the three counties paid for the maintenance of prisoners in penitentiaries the sum of $4,600.28, but in 1900 it was $11,551.55, or nearly three fold what it was a decade ago.

The record of crime in the three counties which are the best representation we can get of the anthracite coal fields, is a lamentable one to contemplate and it shows beyond question that in no other section of our State are crimes more prevalent, and nowhere do they take a more brutal form. They are largely attributable to the sensual and brutal elements of human nature and bear witness to the lack of moral restraint in our people. They are subject to outbursts of passion, when respect for man in whom dwells the true "shekinah" is wholly forgotten. We can get a nearer view of these crimes by the study of the criminal dockets of justices of the peace of typical mining towns, and the impression made by this general study of the three leading counties in the anthracite coal fields will be deepened.

The Dockets of the Justices of the Peace.

The study of the dockets of justices of the peace and burgesses in our boroughs does not tend to exalt one's respect for the courts, in which the first step is taken in the course prescribed for the adjustment of difficulties arising from the frailties of human nature. Some burgesses and justices of the peace are intelligent and well qualified to discharge the duties of the office they hold. They are versed in the law governing their duties; they have a right idea of the importance of just treatment of those who appear before them, and their dockets are well kept. Others are in offices when they ought to be behind prison bars. They know nothing of the law and feel no responsibility in the discharge of their duties. They are politicians and are in the office for "what's in it." The ignorant and innocent are their victims and their dockets are a mass of unintelligent scroll, filthy and torn, and no itemized account kept of the charges in the cases tried by them. When business is dull and these vampires "run dry," they are known to enter ficticious suits on their dockets, return the same to court and collect the fees. Constables are parties to the crime, and are of the same type as the so-called dispensers of justice. It is a coterie of vicious idlers, too lazy to work, too corrupt to be trusted by the average citizen in matters of justice, and too drunk to even enter the few cases they try when a poor victim has been trapped by an unscrupulous pawn. These scoundrels very charily permit the investigator to examine their dockets, and make all possible excuses when asked to produce them. They also have a lawyer near at hand—a man of the same type as themselves, who reminds one of a cunning watch-dog, trained and fattened to safeguard the devious paths of a blundering and arrogant public servant, who owes his position to his trickery as a ward politician. It is a sad reflection on any community to elect such men to offices established for the purpose of dispensing justice in townships and boroughs, but it is only another outcrop of the accursed system of politics whose basic principle is "to the victors belong the spoil."

The local courts of mining towns are kept busy and the volume of business is great. In each borough there are two justices of the peace and, as a rule, they live well on the income of the office. The town of Shenandoah in 1901 sent in 30 per cent. of the criminal cases returned in Schuylkill county although its percentage of population is only 12. The records for the years 1895 and 1896 show that about 50 per cent. of the criminal cases returned to court originated in the above mining town. This is accounted for by the mixed population found in Shenandoah, and the restlessness of the Sclav groups, which represent various nations and tribes, between whom are ceaseless jealousies and rivalries.

The following table gives the business done by justices of the peace in three boroughs located in the Southern coal fields for the year 1900.

Place.	Total No. of Cases.	Disposition of Cases.		Total Cost of Suits.	Av. Cost per Suit.
Shenandoah. Population 20,000.	639	Dismissed Settled Returned Sentenced	22 or 3.4% 340 " 53.3 265 " 41.4 12 " 1.9	$4,412.90	$6.90
Mahanoy City. Population 13,500.	613	Dismissed Settled Returned Sentenced	126 " 21.6 283 " 46.2 204 " 32.2 0 "	3,534.89	5.61
Mt. Carmel. Population 13,100.	586	Dismissed Settled Returned Sentenced	336 " 57.3 120 " 20.5 99 " 16.8 31 " 5.4	3,161.26	5.39

The following table gives the record of three towns in the Northern coal fields for the same year.

Place.	Total No. of Cases.	Disposition of Cases.		Total Cost of Suits.	Av. Cost per Suit.
Olyphant. Population 6,100.	355	Dismissed Settled Returned Def't. not found	159 or 44.8% 41 " 11.5 139 " 39.2 16 " 4.4	$2,213.09	$6.23
Nanticoke. Population 12,100.	295	Dismissed Settled Returned Sentenced	91 " 30.8 39 " 13.2 95 " 32.2 70 " 23.8	1,411.80	4.84
Edwardsville. Population 5,100.	125	Dismissed Settled Returned Sentenced	40 " 32.0 27 " 21.6 45 " 36.0 13 " 10.4	581.25	4.65

Various considerations enter into these accounts which explain the variation in the percentage of crime. In Luzerne county, the judges have called the justices of the peace to account for their lax way of conducting court, and some of the most unscrupulous have been prosecuted. This has had a salutary effect and the number of cases returned has been materially reduced. Then again in the towns of Edwardsville and Nanticoke the burgesses tried more cases than did the justices of the peace. In Olyphant, on the other hand, it is an exception for a burgess to try a case; the justices of the peace try all the cases. In Nanticoke and Shenandoah, we did not succeed in getting all the cases tried, while in the former town a six months' strike had occurred in one of the largest collieries and one of the justices said, "Business used to be twice as good as it is now; the people have no money to go to litigation." From a careful calculation of the number of cases tried before justices of the peace and burgesses in the above-mentioned towns we found that, under normal industrial conditions, there was one case entered for every 20 members of the population, which means, when we consider that there are two persons implicated in each suit, that one person out of every ten of our population was involved in a law-suit of some kind. The number of law-suits instituted in the United States for the year 1900 was one to every 22 of the population, which was also the number recorded in Great Britain during the same year. The costs of suits in local courts ranges from $3 to $6. If we take an average of $4.50 there would be spent annually in litigation by the mining population of our territory about $100,000. Most of this is spent by the Sclavs and, in view of this inexcusable waste, there is no wonder that their spiritual leaders have taken steps to protect their flocks from the political wolves clothed with a brief authority that grow fat upon the ignorance and frailties of these immigrants. It is also a hopeful sign that the Miners' Union has brought Sclav and Saxon closer together, so that they know each other better and the result is fewer suits of law.

The nationality of the persons charged with crime may be

judged from the following table, which is as accurate as the knowledge of the justices of the peace who gave the information. We give also the percentage of population belonging to the Sclav and Anglo-Saxon.

Place.	Percentage of Sclav Criminals.	Percentage of Sclav Population.	Excess of Crime.
Nanticoke	63.5%	50%	13.5%
Edwardsville	66.3	45	21.3
Olyphant	58.8	30	28.8
Mahanoy City	67.3	40	27.3
Shenandoah	79.6	60	19.6
Mt. Carmel	64.8	45	19.8
Average	69.2	45	21.7

This excess of crime among the Sclavs is explained by their social condition and the excess of males over females among them. In the chapter on population we found that there were over 30,000 more males than females among the foreign born in the anthracite coal fields. This excess of males is largely found among Sclavs. Young men with strong natures, far away from the wholesome influence of home life and surrounded by saloons, are liable to break forth into excesses which lead to crimes and their consequences — litigation and imprisonment. It is also true that the offices of justices of the peace are almost wholly in the hands of the English-speaking, and hitherto the Anglo-Saxon has not spared the Sclav the vexation and expense of a lawsuit. This class of our citizens, however, are organizing themselves into clubs and are a power at the polls, so that the political office-grabbers are more considerate than was their wont in the treatment of the Sclav.

As to the nature of the crimes, we have in mining towns in the Northern coal fields 67.5 per cent. against persons and 32.5 against property; and in the Southern coal fields we have 66.1 per cent. against persons and 33.9 against property. In each of the mining towns examined, assault and battery led the list of crimes and formed on an average 50 per cent. of the whole. In the town of Shamokin, in the summer of 1901, three husbands murdered their wives and two others attempted murder. One of the judges of Northumberland county said

Miners Going to Work. (About to descend the stope.)

that, in the last 20 years, 96 murders were committed and only one of the murderers was hanged. In the first 20 years of the existence of Lackawanna county, notwithstanding 106 murders were committed, all the perpetrators save two escaped the gallows. In Schuylkill county since 1880, 84 persons were charged with murder and 6 convicted, while 50 other murderers escaped the vigilance of the officers.* Murderers among Sclavs frequently make their escape. Pay-night in a mining camp where Sclavs live is generally the occasion for dissipation; over their cups they fight and when one of the number is killed, the assailant makes good his escape before the authorities are notified of the murder. When the brute is aroused in the Sclav by a mixture of whisky and lager, he becomes more fierce than the red-toothed beast of prey. The Sclav, however, is improving. A justice of the peace, who had been in office for 10 years in a town where they formed the majority of the population, said that crime is perceptibly diminishing among them. They feel the touch of a higher civilization. More homes are established by them; their priests are better versed in the laws and institutions of our country than they were 10 or 15 years ago; there is less crowding among the bachelors; a larger number of them annually becomes citizens and property holders; and the public school trains their children in a higher life than that in which their parents grew in the fatherland. The wrongs inflicted upon the Sclavs in these local courts long prevailed, and many were the hard lessons taught them by the injustice of avaricious justices of the peace. A better day has dawned and we all look forward to a better type of manhood and womanhood among the Sclavs as they realize more and more the responsibility of citizenship and more fully perceive the blessings they enjoy in a government for the people and by the people.

The English-speaking portion of our population is also given

* The officers of Schuylkill are at present offering rewards amounting to $5,000 for the apprehension of persons guilty of murder. A record of 19 murders in 17 months ought to move the government to vigorous action. In the last two years and a half no less than 76 murders have been committed in the three counties, Lackawanna, Luzerne and Schuylkill.

to crime. One of the results incident to the change in population in these coal fields is that many of the best citizens of mining towns migrate, so that in proportion to the number left the thriftless and dissipated form a larger percentage of the Anglo-Saxon population than in former years. This shiftless class stays here and multiplies, and many of them are deeper in licentiousness and vice than the lower stratum of the Sclav. Of 963 English-speaking criminals in the towns above mentioned during the year 1900, 449, or 46.6 per cent., was credited to the English and American; 105, or 10.8 per cent., to the German; 276, or 28.6 per cent., to the Irish; and 133, or 14 per cent., to the Welsh. We have no accurate means of ascertaining the percentage of the population which each of these peoples forms in the towns specified. As far as observation goes the percentage of crime among the English-speaking nations is about equal.

The Anglo-Saxon and German have a great advantage over the Sclav and Italian. All the varied enjoyments devised to break the monotony of daily life are open to them, while most of them have been born and raised in our country, and the advantages, pleasures and diversities of a higher life invite them to participate of its richness. All this is closed to the Sclav. He is in a new country and among a strange people of a strange tongue. He is the child of a lower civilization than ours, and has grown in an environment not so favorable to the development of manhood as ours. It is not strange the Sclav should seek diversion from arduous toil and the routine of a narrow and monotonous life by frequenting the saloon and diversions which are repulsive to a people of a higher social status. But the Sclav is on the way to a higher and broader life, and the strong physique, the thrift, the reverence for law and amenability to discipline of these people is a promising soil to plant the seeds of a higher civilization.

Crime Among Farmers and Miners — A Comparison.

A family, which moved from a rural community in 1901 and settled in one of our mining towns, expressed great astonishment at the lax ideas of propriety in the relation of the sexes that prevailed; the vulgarity common in the conduct of those whom they met; the profanity and drunkenness heard and seen on the streets, and the low standard of morality and decency prevailing in the town where now they live. They felt thankful that they had no young children to be raised in "such a town as this." They longed for the day when they could again dwell in the quiet country village where the sense of propriety and decorum attained a higher level than in the mining town. There is a difference between the two communities, and possibly the best expression of this is the docket of a justice of the peace of a rural town as compared with that of a mining town. For the purpose of comparison we consulted the dockets of the town of Norwich * in the State of New York, which, taking in the scattered population in the surrounding country, approaches Olyphant in the number of persons residing there.

The population of Norwich is of American stock and homogeneous. In the fall of the year many people come to this town to work during the hop season. Also, when industrial depression invades the eastern section of our State, the number of vagrants invading country villages largely increases and the number of arrests for vagrancy is larger. In the fall of 1897 the industries of northeastern Pennsylvania had not revived from the depression which paralyzed the business of the country for some years previous. In 1900 we were completely over the crisis and all willing hands could find employment. In the fall of 1897 there were 158 arrests for vagrancy in Norwich and in 1900 only 24. The large number of arrests for vagrancy in the former year swelled the number of crimes to 384, while in 1900 the number was 233. Vagrancy cannot be laid to the homogeneous population of Norwich, so we leave it out of the count. In mining towns this crime is seldom met with. Out of over 2,830 cases investigated in mining towns we did not meet one for vagrancy. With this elimination we have then in Norwich 226 crimes in 1897 and 209 in 1900, or an average during these two years of one crime for every 32 persons. Of

* In the last census Norwich town, including village, has a population of 7.004.

the 226 and 209 crimes in the years specified, 162 and 154 were arrests for drunkenness. In the 2,830 cases read on the dockets of justices of the peace in mining towns we only found one of them attributed to drunkenness. To get drunk is no crime in mining towns. If it were, many of our burgesses, policemen and justices of the peace would be frequently chargeable. It is a frequent sight to see men in broad daylight on a Sabbath day drunk on our streets, and the mothers, for diversion, bring their children to see the antics of the inebriate as he goes home. There is not so much drunkenness in a rural village as in a mining town, and yet 162 and 154 are arrested for drunkenness in the former and virtually none in the latter. Suppose we eliminate these again from the crimes in a rural district, then we have in 1897, 64 offenses and in 1900, 55 or an average of 8.5 per 1,000 population.

The cases of drunkenness referred to in Norwich took place for the greater part in the fall of the year; the season when farmers bring in their harvests "rejoicing." Once a year these sons of the soil dissipate and have a "good time." There is periodicity observed also in the drink habit of mine employees, but it is every pay-day with them and since the two weeks' pay a large number of men have a "good time" twice as often as formerly.

Of the crimes recorded on the docket of a rural town drunkenness forms 72 per cent. and assault and battery 9 per cent. In mining towns drunkenness is out of the count and assault and battery forms 58.3 per cent. Almost all of these assaults are committed when the passions are inflamed by spirituous liquors.* The list of weapons used in committing the assaults

* The percentage of crime due to drink in these coal fields has been variously estimated. Col. H. M. Boise has placed it at 75 per cent. Others place it at 70 per cent. In "Economic Aspects of the Liquor Problem" we read: "As a final result we learn that intemperance was on the average a cause of crime in 49.95 per cent. of 13,402 convicts, while as a principal cause intemperance occurred on the average in 31.18 per cent. of all cases." Intemperance as a cause of crime can be best studied in local courts and to take the percentage of crime due to this cause from the returns of penitentiaries is misleading if we want to find the social effect of intemperance as a factor in crime. Professional criminals are temperate, but the men who

by persons in mining towns shows how devoid they are of self-control. Among the implements mentioned are, stone, knife, revolver, razor, bar, sprag, chain, dinner-pail, axe, lead-pipe, cuspidor, hammer, chisel, bolt, hatchet, pick, shovel, etc. Anything the hand can lay hold of in the excitement of the moment is used, and when no object is available they use their teeth and boots. Accounts of some of the fights among the Sclavs show them to be so brutal and savage that the sensitive is sickened when they are rehearsed. Men are not the only guilty ones; females, when they are enraged, are also found as fierce and blood-thirsty as the men. Many women are brought before the local courts charged with threats: some of the cases are: threat to break the leg of her neighbor; threat to strike out her neighbor's teeth; to drive a knife through her neighbor; to blow up her neighbor's house with dynamite, etc. Women and mothers here threaten to commit outrages in words which would shock a rural community even if the most confirmed tough were to utter them. The dockets of mining towns, on almost every page, are repellent because of deeds due to unrestrained passion, indulgence of vicious propensities and outbursts of brutal savagery. The story is very different while perusing the criminal docket of a rural justice of the peace.

The comparison of the disposition of the cases is also very suggestive. The following table shows the disposition made by the justice of the peace in Norwich:

Disposition.	1897.	1900.
Sentences Suspended	104 or 27.08%	79 or 33.90%
Discharged	160 or 41.66	51 or 21.88
Sentenced	112 or 29.16	99 or 42.49
Returned	8 or 2.10	4 or 1.73
Total	384 or 100%	233 or 100%

The average returned to court from this office in the above

quarrel in their cups fight, and, although two or three lawsuits may be the result, yet, when the heat of conflict is over, the men will settle the quarrel and not go to court. In mining towns the percentage of crime due to drink is between 55 and 60. The variation is due to the nature of the population. In towns where the Sclavs form a considerable portion of the population it is higher than in towns where the English-speaking inhabitants preponderate.

years was 1.91 per cent. of all the cases tried, but the average percentage returned to court from the offices of the justices of the peace in mining towns was 48.44. The suspended sentence, so often exercised by the New York justice of the peace, is unknown in mining towns. We believe if it were exercised by our peace officials that the effect would be far better upon many of these weak men than a commitment to prison to await trial or to discharge a sentence. The rural justice of the peace is kept busy during the fall of the year, but at other seasons he has little to do. Those in mining towns have about the same number of cases the year round, month after month bringing its unfailing crop of deeds of violence as regularly as the mine employees receive their pay.

The above comparison explains in part the difference felt by the family above referred to when it moved into one of our towns from an agricultural village. If this difference is to be attributed to one cause more than another, it is to the drink habit, which destroys moral fiber, blunts the moral sensibility and lets loose upon society the animal propensities. This is not the only cause, however. The occupation of digging coal is dirty and disagreeable. Its effect is everywhere seen, and although the workers wash ever so clean, anyone familiar with miners will always know them though dressed in their "Sunday clothes." They inevitably lead a coarse life and although thousands among our mine employees are as well versed in the qualities of gentleness, veracity, honor, decency and reverence as any class of working men in any part of the country, yet another large portion of them live in associations that are vicious, habits that are demoralizing, conversation that is defiling, and a moral standard that is far from the level of that chalked out by the Christian conscience.

Our Boys and Girls Who Go Wrong.

At the close of the year 1900 there were 1,553 juvenile delinquents in the reformatories of the State, not counting the 513 in the Huntingdon Industrial Reformatory. The average age of the 1,553 young transgressors was 14.6 years. The 780

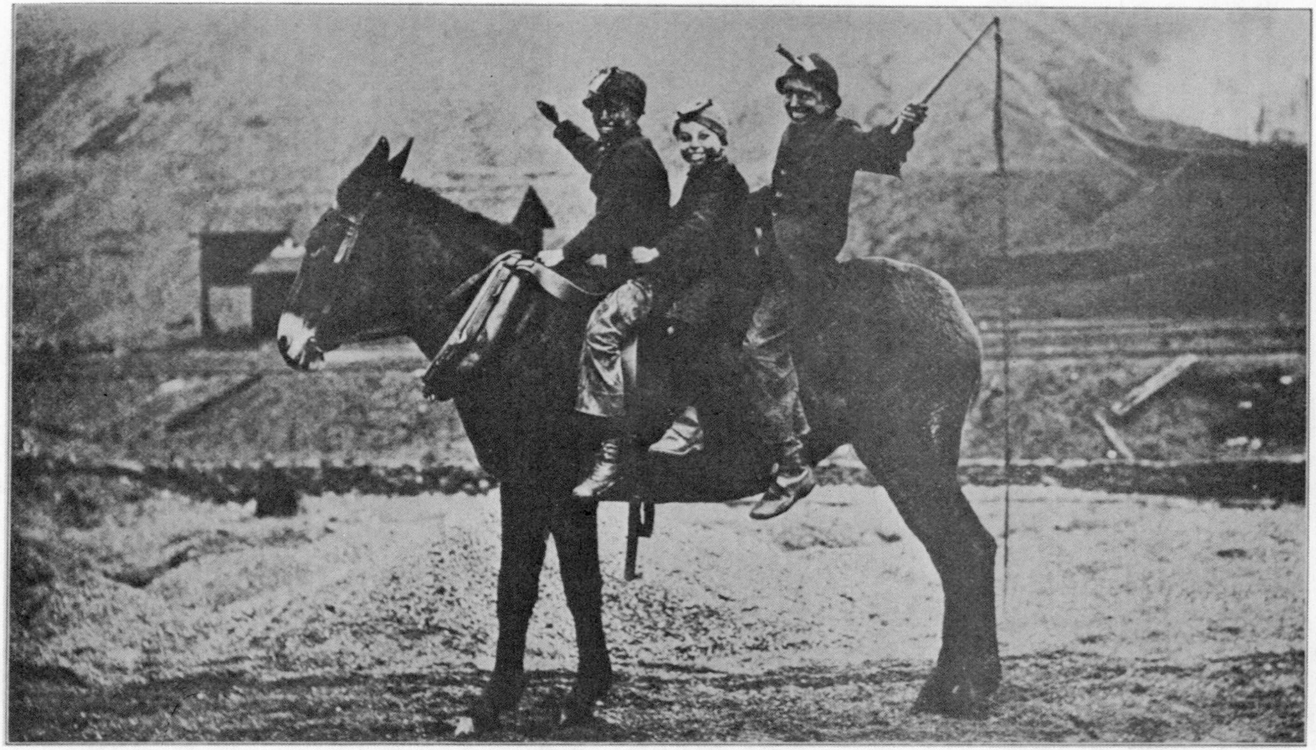

THE WINNING BOYS.

committed to these institutions in the year 1900 are classified according to ages as follows:

Ages.	Number.
Between 7 and 9 years,	12
" 9 and 11 "	79
" 11 and 13 "	168
" 13 and 15 "	250
" 15 and 20 "	271

Of these young criminals 103 or 13.2 per cent. was foreign born, 669 or 85.8 per cent. native born and there were 8 whose birthplace was unknown. In the year 1890 the moneys spent by the State upon the Huntingdon Industrial Reformatory, the Home of Refuge and the Reform School amounted to $188,-534, while the expenditures in 1900 amounted to $479,598.19. The total of young criminals in the reformatories of the State was 2,066 or 340 per million population, and each of these costs the State an average of $242 per annum. Of the youths in our State between 10 and 20 years, 2.06 per 1,000 are in State Reformatories kept at public expense. Beside these institutions there are many others maintained by private corporations and individuals, where the incorrigible and delinquent are kept.

In the anthracite coal fields there are the "Home for Children" in Pottsville, the "Home of the Good Shepherd" and the "Home of the Friendless" in Wilkesbarre, and in Scranton the "House of the Good Shepherd," the "Home of the Friendless," "St. Joseph's Foundling Asylum," and the "St. Patrick's Orphanage Asylum." In these institutions there were 762 persons in January, 1901, and the sum of $51,-575.22 spent in their maintenance for the year 1900. They are the homeless, dependent and delinquent, although we cannot say how many of them belong to the latter class. Some juvenile delinquents are also kept in county jails, a practice against which the Secretary of Public Charities has protested, calling it criminal, and thrice so is it when these boys are not separated from adult prisoners.

In the year 1900 there were in the reformatories of the State 122 juvenile delinquents from the three counties, Lackawanna,

Luzerne and Schuylkill. Of these 88 were boys and 32 girls, and they cost nearly $22,000 for maintenance.

It is very difficult to get the exact number of juvenile delinquents in our territory for the reason that benevolent institutions intervene whose agents take charge of the children. But those who have resided for some years in these coal fields and have given attention to this question know that this class of transgressors is large. In one of our towns a few years ago three boys, from 12 to 16 years of age, started fires in four places on the same night. The wind was high and they selected the location for three starts out of four, where, if the prompt action of the fire companies had not extinguished the flames the whole town would have been reduced to ashes. They succeeded, however, in destroying that night property valued at $50,000. In another town a young girl of 15 years of age was reprimanded by her father for walking the streets at two and three o'clock in the morning. The following day she took carbolic acid and left a note wishing her friends good-bye and saying "she was going to Jesus." In another town a gang of boys, regularly organized for criminal purposes, committed many robberies and for months succeeded in eluding the officers. The climax was reached, when in broad daylight, they held up a young lad and robbed him of the pay he had just drawn. Their retreat was traced to an old working in an abandoned mine in close proximity to the town, where they stowed away many of the articles stolen. In another town three boys between 14 and 16 years were arrested for attempted rape under circumstances which reminded one of the savagery of primitive people. Three other boys were recently arrested charged with robbery and attempted murder. The boys of one of our breakers were promised a sleigh-ride by the outside superintendent of the colliery. The date was fixed but a thaw came and the pleasure had to be postponed. The boys got angry, upset the company's wagons, took their sleighs away a distance of two miles and the next morning went on strike. Three boys, from 7 to 11 years of age, in one of our cities, held up, at the point of a revolver, a young girl and robbed her of her pay. One of the most notorious criminals in the Eastern States, whose picture is in the rogue's gallery, served his apprenticeship in crime in one of our towns, and recently the authorities traced a gang of thieves which had committed many depredations in mining towns of the anthracite coal fields to the very town in which this desperate character was born.

Bastardy and drunkenness are frequent. In a mining camp of not over 400 population, five young girls gave birth to children out of wedlock in the last three years. Ten girls are in trouble in one of our towns at present as the result of the soldiers' encampment of the summer of 1902. On summer evenings it is nothing unusual to find young girls parading the lonely walks near mining towns until past midnight. In some instances their shameless conduct reminds one of conditions which must have prevailed at Sodom and Gomorrah, or the practices which prevailed in the groves surrounding the temple of Aphrodite. In all our towns there are "chippies" and some saloons open their doors to traffickers in lewdness. When one of these young girls was asked if she was not afraid, her reply was: "Not as long as the drug store is handy." Men in a position to know say that many of our young boys are given to practices which undermine their health. Two boys were seen perpetrating onanism on the streets of one of our towns. Doctors and druggists frequently have patients who are victims of this vice and who find relief in early marriage.

Many of these evils arise from the want of moral sensitiveness in the homes in which these boys are raised. The relation of the sexes is not what it ought to be. The prospect of marriage is frequently made to excuse licentiousness. One of the pastors of our towns said: "In most of my weddings I marry three and not two." The moral and religious training of youths raised in mining patches is woefully neglected and the consequence is the irregularities which exist. Others raised in idleness and unable to get positions to suit them out of the mines drift into crime. The boys are not taught trades for there are no facilities for such teaching. We have no industrial schools. If a boy commits a crime and is sent to the Industrial Reformatory, he has a chance to learn a trade, but if he is a good boy and stays at home the chances are 19 to 1 that he will be compelled to enter the mines for a livelihood.

A man of considerable experience in ferreting out crime once said: "The greatest curse of these towns is the slot machine." A close watch on pay-night of one of the stores where these gambling devices are located will confirm the above statement. Boys from ten to twenty years of age, when they have money, cluster around these and, infatuated with the whirl and click of the machine, they stand there until the last penny is spent, and then curse the machine. They will steal from the home, take money from their pay, borrow from their friends, and devise cunning schemes whereby they may get the means to carry on the game. All their mind is concentrated upon it. They calculate and watch the revolutions of the machine with the intensity of the hound after its prey. If they win their joy is great, but though they win they leave not. There they stay staking again the gain in further chances until at length every penny is spent and they are "dead broke." Nothing in these towns fosters the gambling tendency as do these slot machines. They are the greatest breeders of crime known in these communities.

The amount of money spent on these, as in every other business, varies. One saloonist, pointing to a machine, said: "That brought me $35 yesterday." That was the maximum he ever realized in one day. An ordinary pay-day would net him from $8 to $12. The gains depend partly upon the nature of the machine and the wealth of the players. Boys on an average spend from 50 cents to a dollar each pay-day; young men spend from $1 to $3. Men infatuated with the machine are known to spend $5 in chances, while a foolhardy Pole "broke" the record, as far as we have heard — he stood before the machine all day and spent $30.

Among the patrons of these machines are found representatives of all nationalities. The Irish seem to be in the lead while the Poles show a strong tendency to games of chance.

A man who ran four of them said: "Young America is very fond of the slot machine." Those who play, as a rule, do not indulge in intoxicants. When machines were kept in saloons, it was nothing unusual to see a company of young men going from saloon to saloon to play at the machines, and still not touching drink. These young players are superstitious. Almost every one of them has his favorite machine; on that he will stake and on no other. When players lose, it is curious how they personify the machine or believe it influenced by powers which bring them bad luck. Some Sclavs curse the machine when they have lost much and threaten to smash the idol. Others look upon the machine as a pure piece of mechanism which can be manipulated to order. Some of these boys "beat" the machine. One of them manipulated it with a wire and stopped the wheel each time at a point where returns were secured. Another proprietor opened one of his machines after pay-day and found among 800 pennies about 400 washers. These the boys had purchased in a hardware store, three dozen for five cents, and found them "to work like a charm."

Machines, when first introduced into the Wyoming and Lackawanna valleys, were, for the greater part, managed by a corporation. They were placed in saloons and the profits equally divided between the owner of the machine and the saloonist. It was a paying business, and soon many saloonists bought machines of their own and secured all the profits. It is claimed that from 40 to 60 per cent. of the money staked is profit.

The slot machines in saloons, however, soon became so great an evil that the courts of the several counties insisted upon their removal. The crusade began about three years ago. In Schuylkill County each saloon keeper has to swear, when he applies for license, that he has no slot machine on the premises. In Lackawanna and Luzerne counties the constables in making their returns are placed under oath that, as far as their knowledge goes, no slot machine is in any saloon within their territory. This crusade instituted by the courts has resulted in banishing the machines from public houses, but they are still found in cigar and tobacco stores, in candy and barber shops, etc. There are some machines still kept in saloons, but they are secreted and only for the use of private parties. In many of the mining towns in the Northern and Middle coal fields the click of the machine may still be heard in saloons.

In candy and barber shops, cigar and tobacco stores, the machines are operated without interference. This is in violation of state law and it is difficult to see why the courts permit what they regarded as an evil in saloons, to run unmolested when the location is changed. The stern hand of the law should remove this source of corruption in mining towns. The slot machine is the greatest curse now existing among our youths. Those in charge of the machines observe no days or seasons. The wheel goes on Sunday as well as on week days and, in many instances, the first day of the week is the favorite of the players. While the officers of the law slumber the gambling evil grows. In a picnic held last summer in a mining town of 13,000 population, no less than $570 were staked on the wheel.

There is very little prize-fighting or dog-fighting in our mining towns. Those inclined to sport find their chief diversion in quoits, shooting matches, and in cock-fighting. These are chiefly patronized by youths of Anglo-Saxon extraction, and are invariably associated with certain saloons whose proprietors are the leaders in the sport. The stakes in quoit matches generally go up to about $25. Shooting matches are arranged for $100 to $200. The stakes in cock-fighting seldom pass over $50. The number witnessing the sport depends upon the popularity of the persons interested. In shooting matches a crowd of several hundred assembles. When a "main of cocks" is pitted the number present depends upon the capacity of the room and the publicity of the affair. In boroughs, cock-fighting is generally kept secret, for the constables interfere. Some saloons, however, have cock-pits in their cellars where matches are frequently carried out. As a rule, matches are arranged in mining towns outside borough limits. In a town of Schuylkill county, a company of these men bought an old school-house for the expressed purpose of cock-fighting and, from November until

April, hardly a week passes but a match or two takes place, where on an average about 200 persons are assembled.

All ages follow these contests. Young boys of 13 side by side with aged men may be seen there. In none of these sports is the passion for gambling stirred as in cock-fighting. Quoits and pigeon-shooting are tame beside it. Around the pit young boys and old men gather and become roused to a pitch of excitement seldom seen outside halls where prize-fights occur between well-matched contestants. Much betting takes place: the bets range from ten cents to as many dollars. Boys and young men, given to this sport, are completely carried away and sometimes resort to unnatural means to secure money for betting. One of the most unnatural victims of this passion ever known to us was a young man, who mercilessly beat his mother, because she would not give him $10 to sport in cock-fighting. Another man kept $5 from his pay for this purpose. He had a wife, however, who suspected him of sharp practices. She went to the mine foreman to find out the amount of her husband's pay and followed him to where the cock-fighting was carried on. She brought him home and the neighbors say it was the last he attended.

The winning boy appeals to all men, and it behooves society to put forth every possible effort to save him. Mayo-Smith has said: "Our statistics seem to show that the second generation, that is, the native whites of foreign parentage, are peculiarly subject to deteriorating influences." Anyone familiar with these coal fields knows that to be true. The children of Anglo-Saxon immigrants are not so honest, so frugal, so moral as their parents were, and the same tendency to degeneracy is observed in the native born of Sclav parents. Hereditary traits do not account for youthful criminals. They are rather the product of bad bringing up. It has been observed that children who become criminals have been raised in homes where proper restraint and moral training were rare. Many of these children lose both or one of their parents before they are sixteen years of age. The youthful delinquent may have inherited more than the ordinary human being's share of mental dullness and incapacity, but the one factor above all else which accounts for his conduct is bad training, and where good training has been substituted it has in most cases brought forth the fruits of righteousness.

In the year 1901 the Legislature of our State passed the Juvenile Court Act, which commits to the judges of the Court of Quarter Sessions discretionary powers relative to juvenile dependents and delinquents tried before them, and authorizes them, if they think it best for the future of the child, to put him in charge of a probation officer. For the first ten months the act was in operation, the judges were called upon to deal with 366 dependent children and 739 delinquents in the city of Philadelphia alone. This shows the great need of some such effort. Last February, however, this laudable attempt at dealing with juvenile law-breakers was declared unconstitutional by the Supreme Court of the State. It was a conscious effort on the part of philanthropists to prevent delinquents from becoming criminals, by placing the erring one, who is too young yet to fully realize his crime, under the restraining and guiding hand of an officer of the court. The good done by Juvenile Courts has been testified to by many public-spirited citizens and the State of Pennsylvania ought to get the benefit of such an act. Miss Minnie F. Low in a recent address said that in the city of Chicago there were "575 boys under the age of 16 years lodged in the county jail each year awaiting the action of the grand jury." But in the last three years "since the establishing of the Juvenile Court, instead of 575 we have averaged less than 20 boys held to the criminal courts annually." That record ought to commend the Juvenile Court and the probation officer to every State in the Union, and the Keystone State cannot show justice and mercy to the thousands of delinquents within its borders unless some such system of dealing with them is adopted.* The delinquents invariably come from among the poor whose parents

*Laws, drafted by competent jurists, were passed by the last Legislature, providing for Juvenile Courts and Probation Officers, which, all those interested in juvenile delinquents hope, will meet all requirements.

have not the means to make good the evil wrought by the youth. To imprison the boy, as is now done in our counties, is a crime. What we need is the restraining influence of a prudent person, to whose watchful care the boy is committed, who can advise him in a kind spirit and still have behind him the power of the law to command respect. Every youth deserves the protection necessary to make him a useful member of society. Respect, honesty, and reverence are the factors needed to keep these boys in the right way.

CHAPTER XI.

"CHARITY NEVER FAILETH."

1. THE LIBERALITY OF DIRECTORS OF POOR RELIEF. 2. HOW MUCH DO THE COAL COMPANIES GIVE? 3. INDIVIDUAL GIVERS.

THE LIBERALITY OF DIRECTORS OF POOR RELIEF.

In the anthracite coal fields there are ten districts in which the civil authorities dispense aid to dependents or furnish a home to them where they are wholly kept by the taxpayers of the district. In order to give an idea of the number of dependents in this territory we give the following table compiled partly from the report of the State Board of Charities and partly from personal investigation:

Poor District.	Number Relieved Out-Door.	Per 1,000 Population Relieved.	Number in Almshouse.	Per 1,000 Population in Almshouse.	No. of Widows and Children Relieved.	Per 1,000 Population of Widows and Children Relieved.	Per Capita Expenditure in Poor Relief.	Percentage Greater or Less than Average of State.
State.	33,619	5.3	12,168	1.9	28,754	4.5	33.7 c.	
Schuylkill	1,273	7.3	310	1.8	501	2.9	47.9c.	+42.1%
Coal township	700	22.8	44	1.4	654	21.3	112.8	+234.7
Middle coal field	2,135	20.6	206	2.3	1,200	13.3	40.8	+21.1
Central	2,466(?)	17.2(?)	186	1.3	1,800(?)	12.5	47.4	+40.6
Ransom	850	18.6	100	2.2	650	14.2	31.0	— 8
Scranton	1,611	14	221	1.9	674	5.9	47.6	+41.2
Blakely	334	12.2	22	0.8	156	5.9	33.3	— .1
Carbondale	200	14.7	16	1.2	160	11.8	35.1	+ 4.1
Average		15.9		1.6		10.9	49.5	+46.9

The following table gives the number in almshouses, the amount per capita of population expended in maintaining them, and the percentage decrease as compared with the average expenditure of the State, in three agricultural counties and in two bituminous ones:

County.	No. in Almshouses.	Per 1,000 Population.	Per Capita Expense.	Per Cent. as Compared with State.
Bucks (Agr.).........	142	1.9	21.8 c.	—35.2
Berks "	189	1.9	17.1	—49.3
Chester "	226	1.4	29.5	—12.5
Clearfield (Bit.)...	105	1.3	31.2	— 7.5
Westmoreland " ...	185	1.1	22.7	—31.7

The following table gives us a comparison as to the expenses incurred in the maintenance of almshouses and of out-door relief, as estimated in per capita of population:

	Out-door Expense per Capita of Population.	In-door Maintenance per Capita of Population.
Average for State..............	6.8 c.	26.9 c.
In anthracite communities...	19.9	28.8
In bituminous "	8.7	18.2
In agricultural "	2.8	20.0

From these tables we see that the number per 1,000 population receiving out-door relief in anthracite communities is unusually high — almost three-fold the general average of the State. If we compare the number with that of older countries it is not excessive. England has an average of 19.5 persons per 1,000 receiving out-door relief. But the anthracite coal fields have neither the economic nor the social conditions of older countries. The majority of the mine employees and those dependent upon them are in the prime of life and in the years of greatest productivity. We have not the numbers usually found in normal communities in the higher age groups. The large number of widows and children receiving out-door relief accounts for the high average, and it brings forth very clearly the heavy burden imposed upon these communities because of the large number of fathers and husbands killed in this industry. But making due allowance for the killed and the maimed in anthracite collieries, there is no reason why the number of persons receiving out-door relief should be as large as it is. The per capita sum spent for this purpose is over three times that of the State; over twice that of the bituminous fields, and eight times that of agricultural counties. There is considerable difference in the districts. Coal township stands at the head of the list and Schuylkill county at the bottom. Three years ago the latter had about the same proportion as the former, but in the last few years the Tax-payers' Association of Schuylkill county took the list of out-door relief in hand and thoroughly purged it of its abuses, and, without working injury to the worthy poor, succeeded in reducing the number 50 per cent. and the expenditures were cut down from $40,000 to $25,000. The abuses unearthed by this band of business men were amusing though serious. Some persons dead for over two years still continued to draw relief, while many "respectable" people surreptitiously took relief while from $90 to $100 came into the house each month. It is the conviction of those who have studied this question that the lists in most of our districts could, without working any wrong, be cut down 50 per cent. The following quotation from the report of the Tax-payers' Association of Schuylkill county reflects the gross neglect on the part of "poor directors": "The directors of the poor in April, 1903, intend to suspend all orders now in force and request new applications to be made. This has not been done for ten years and is very necessary in the condition of the present lists. This cannot but help in weeding out parties who are now recipients who do not deserve relief."

The above tables also show that the abuse of out-door relief is greater in the Middle and the Southern coal fields than in the Northern. The three first districts given in the above table are in the former and the remainder in the latter section of our territory. The districts of Scranton and Central Luzerne have, in connection with the almshouse, an insane department where defectives are kept. The expenditure in each of these almshouses, computed in per capita cost of maintenance, is about the same as that of the Schuylkill district where also the insane are kept. Schuylkill county in its poor rate, showed the same abuse as does Coal township and the Middle Coal Field District, before the Tax-payers' Association was organized and made itself felt. In many of the mining townships in Schuylkill county the coal companies pay from 85 to 95 per cent. of the taxes, and the extravagance in the disbursement of poor relief is partly accounted for by the fact that the electors of these townships do not pay the poor tax. The heavy burden of taxation which fell upon the coal companies in Schuylkill county was the reason that the Tax-payers' Association was organized. The root of evil is politics. The management of Schuylkill Haven almshouse has been a constant source of scandal for the last quarter of a century, and lately one of the daily papers asked : "Will there ever come a time in Schuylkill county's history when the almshouse will be entirely free from political scandal and the Poor Directors be free to act as they deem best ?"

The growth of pauperism and insanity in the anthracite coal fields for the last decade has been marked. For the purpose of comparison we will take the number of paupers and insane in the districts which are wholly in the coal fields. In 1890 there were 887 or 178.3 persons per 100,000 population in the almshouses, and in 1900, 1,514 or 242.6 per 100,000 population. During the decade the percentage increase of population in the counties of the coal fields was 25.4 per cent., but the inmates of almshouses increased 70.6 per cent. In the State in 1890 there were 173.3 persons in the almshouses per 100,000 population, and in 1900, 193.07. The percentage increase of population in the State was 19.9 per cent., but the percentage increase of paupers in almshouses was 33.5 per cent. Pauperism has increased in the State in the last decade, but not to the extent it has in these coal fields. The increase in the expenditures in the last decade was 35.37 per cent. There is a decrease in the per capita cost of maintenance of paupers from $119.38 to $94.67 per annum, but the amount appropriated for salaries increased 33.8 per cent.

The following table gives us the paupers and defectives in the almshouses in the coal fields for the years specified.

	Sane.		Insane.		Blind.		Deaf & Dumb.		Children.	
	M.	F.	M.	F.	M.	F.	M.	F.	Sane.	Insane.
1890	412	219	117	112	13	6	2	2	26	3
1900	530	214	382	325	15	3	0	0	47	6

Of the sane dependents the number of females has not increased but that of the male has 28.63 per cent. The number of insane persons in our institutions has greatly increased. We cannot compute the percentage of increase by the above figures; the hospital for the insane, "Retreat," in the Central district was only opened in 1899, and the insane of the district, who were scattered in various institutions, brought there. Counting, however, all the insane of the three counties in 1890, we have 694, and in 1900, 1,106, which is an increase of 59.36 per cent. The increase in our almshouses then comes chiefly from insanity and all who know the close connection between the increase of alcoholism and of insanity will undoubtedly conclude that they are related as cause and effect in these coal fields. Insanity has increased twice as rapidly as population.

The number of insane in the three counties in 1890 was 134.4 per 100,000 population, and in 1900, 177.2. In the State in 1890 the insane numbered 187.69 per 100,000 population and in 1900, 247.48. If we express this increase in percentages that of the State is 57.93, and that of our counties 59.36, but in the years in which the above comparison is made the proportion of insane is less in the coal fields than in the State as a whole.

In the year 1890 the native born in almshouses was 28.4 per cent., but in 1900 it was 47.5 per cent. Thus during the decade the native born inmates of poorhouses in our territory increased 67.2 per cent., while the foreign born inmates decreased in that period 26.7 per cent. In the almshouses of the State the proportion of native born remained nearly constant during the decade, the increase being only 5 per cent.

The total amount expended in the three counties on paupers and defectives of all classes in 1890 was $190,274.68, and in 1900, $284,779.88, which was an increase of 43.5 per cent. The amount disbursed in out-door relief increased in the decade 75.7 per cent. The greatest percentage of increase took place in Luzerne county. In the three counties in 1890 the per capita expenditure for charitable purposes was 38 cents, and in 1900

45 cents, while in the State it was 33 cents in 1890 and 34 cents in 1900. In the years compared our counties exceed the general average for the State.

The districts have farms connected with their almshouses varying from 90 to 150 acres. The returns from the farms are not great, for the testimony of the managers invariably is that the inmates will not work. The houses built for the accommodation of the inmates are on the average in good condition, save that of Schuylkill Haven, which suffers from what the Secretary of the Board of Charities calls "too much politics." It is to be hoped that the bill * introduced into the Legislature this session will pass, giving the care of the institution to a board of commissioners who shall serve free of charge and have full control of the in-door and out-door relief. This will take it out of the hands of the politicians, and the work of caring for the poor and defective in the county will cease to be a source of public scandal, shameful wastefulness and criminal incapacity.

Besides the appropriations made for charities by municipal taxation, the State annually appropriates $122,175 to the ten hospitals in the anthracite coal fields. This will probably be increased in the current year to $134,175 because of the changing of Lackawanna Hospital, Scranton, from being a private institution to a state hospital.

Hence we see that the total appropriations made for the paupers, defectives and injured by both the municipalities and the State annually amount to over $420,000, not counting the amounts invested in real and personal estate in the various almshouses and hospitals.

How Much do the Coal Companies Give?

Coal companies and other organized bodies, such as charity organizations, benevolent societies and churches, practice charity, but the exact amount contributed by these corporate bodies cannot be obtained. All we can do is describe the various ways in which these bodies practice this virtue.

Coal companies and churches contribute to the maintenance of

* This beneficent bill failed to pass the Legislature of this year.

the hospitals in the anthracite coal fields where the injured and the sick among the working classes are freely treated. We have eleven hospitals, three of which, Ashland Hospital, Hazleton Hospital and Lackawanna Hospital are State institutions. Moses Taylor hospital, in Scranton, is maintained by an endowment fund amply sufficient to meet all expenses, and is exclusively for the injured of the Delaware, Lackawanna and Western Company. The other hospitals in our territory are the Hahnemann of Scranton, the West Side of Scranton, the Wilkesbarre of that city, the Mercy of Wilkesbarre, the Pottsville of that city, the Pittston of that city, the Carbondale of that city. Each of these institutions receives State aid from $2,000 to $16,000, and whatever contributions are made by the coal companies and the churches are to one or the other of these hospitals. These institutions are open to all creeds and nationalities. Of the 2,874 persons treated in five of them in the year 1900–1901, 989 or 34 per cent. was Protestant and 1885 or 66 per cent. was Catholic. Of the 1,998 patients treated in 1898–1900 in Ashland hospital, 793 or 39.6 per cent. was native born and 1,205 or 60.4 per cent. foreign born. Among the patients nearly all European nations are represented. The mine employees form about 80 per cent. of all the persons treated in these hospitals, and of the total number treated over 75 per cent. of the cases is surgical. These hospitals which know no distinction in creed or nationality, have a claim upon corporations which, because of industrial considerations, should keep them in mind.

In an exhibit presented to the Coal Strike Commission in behalf of the coal operators of the Northern coal fields, it was stated that the coal companies and individual coal operators had contributed $25,225 to Lackawanna Hospital. The hospital was founded in 1871, so that during the years of its usefulness the coal companies contributed less than $1,000 a year, while fully 60 per cent. of the patients treated there was of the mining class. During this time the State contributed $232,000 and the public and other institutions over $125,000. So that if we take into consideration the work done by this institution and the years it has existed, the contribution of the coal companies is very meagre.*

The following list of private institutions shows how they were maintained according to the last reports issued by them.

Hospitals.	Total Cost of Maintenance.	Contributed by Coal Operators.	Contributed by Coal Employees.	Contributed by the Public, Etc.	Contributed by the State.
Lackawanna	$28,626	3.8%	2.3%	33.8%	60.1%
Pittston	11,825	None.	15	15	70
Wilkesbarre	26,435	25.5	None.	12.6	61.9
Carbondale	7,941	2.5	"	30	63
West Side (Scranton)	4,144	None.	"	50	50
Pottsville	27,974	"	"	72.8	27.2
Mercy (Wilkesbarre)	13,022	"	"	84.6	15.4
Hahnemann (Scranton)	12,530	"	"	82.6	17.4
Total and average	$132,497	4%	2.2%	47%	45.7%

Thus, barring out Moses Taylor Hospital, the total contributed by the coal companies toward the hospitals in the anthracite coal fields in one year was $5,299. In addition to this some of the companies gave a car of coal, worth about $100 to the Carbondale and Pottsville hospitals. After careful computation, and allowing for what some operators contribute when their injured employees are treated in private hospitals, the total amount contributed by these companies for one year is about $8,000, or an average of 13 cents for every 1,000 tons of coal mined. It may be said, in extenuation of this neglect on the part of coal companies, that the individual operators contribute privately and that most of the funds of the

* The coal companies, as above stated, pay the major part of taxes for the relief of the poor. The following extract from the testimony of G. H. Butler, one of the directors of the Central Poor District, before the "Commission to Inquire into the Condition of the Insane in Hospitals," shows this: "You will understand that there is a great deal of mineral and coal lands in the district and the large corporations pay a large percentage of these taxes, for instance such corporations as the Delaware & Lackawanna, Lehigh Valley-Susquehanna Coal Company, and a number of others, and the men represent, ing these companies are watching and scrutinizing everything that is being done and have a tax-payers' association; they pay seventy-five per cent. of all this and they watch and scrutinize everything and have a system of audit, etc."—Report, 1902.

21

State are collected from corporations. But it is well known that the individual operators residing in the coal fields to-day are very few, and although capitalists do pay largely to the State funds for the privileges they get, yet a larger portion of the wealth taken from these coal fields should be donated to these charitable institutions for the maimed and the sick of the working classes. A gentleman in charge of one of our hospitals went to the operators to ask for aid, which they refused; he then said: "Won't you give us some coal to keep the patients warm while we patch them up for you?" The operator said: "Why, that is the same as money, for we sell coal." After repeated solicitations $500 and 25 tons of coal have been contributed to this institution in which 80 per cent. of the patients treated are of the mining class. The Pennsylvania Railroad Company made a handsome contribution to "Retreat," the new insane hospital erected by the Central Poor District, by expending over $25,000 in grading and improving the grounds where the hospital stands.

Many of the companies contribute something to the families of the killed and injured among their employees. The Philadelphia and Reading had on its lists last November houses whose aggregate monthly rent amounted to $140 which it did not collect. It was its contribution to the widows and indigent among its employees. The Markle Company, the Wentz Company, the Coxe Company as well as other companies in these coal fields, sometimes cancel debts for rent and coal incurred by widows whose husbands have been killed in the mines, while in all our investigations we only found one case where a widow was evicted, and the cause then was that she dared to open a store in a company house to aid her in the conflict of life. The company owned all houses and land in the town, and it does not interfere with widows who wash and scrub to keep the wolf from the door, but they must not open a store. Most of the companies also contribute something for burial expenses and it is well that they do, for medical science in Pennsylvania has its eye on the corpse of the poor. If a poor widow appeals to the Poor Board for help to bury her husband killed in the mines, she is told that, by the law of the State, it is liable to a fine of $500 if it exercises this charity.* All it can do is to relieve her of the remains, send them to Philadelphia, where the body will be used for the advancement of science. It is not enough for young graduates to experiment on the living in these hospitals where the poor among mine employees are treated, but the dead of the poor must also serve the purpose of science. The widow and children of the poor must forfeit the privilege of having a grave where the dead may be laid and a flower placed in loving memory of him who labored and died for them, unless the workmen and the employers contribute the requisite amount to bury the dead. Well is it then that funds and contributions are made that the poor may bury their dead in sacred and consecrated ground, and to the credit of both mine employees and employers be it said, few are the bodies shipped to Philadelphia for the advancement of science from the coal fields.

There are volunteer charity organizations in Scranton, Wilkesbarre, Hazleton, Pottsville and Carbondale. These, for the greater part, coöperate with the poor directors or conduct a work of their own along strictly religious or humanitarian lines. In the cities of Scranton, Wilkesbarre and Pottsville there are five institutions wholly for charitable purposes, where 762 dependents and delinquents are maintained at an annual expense of $51,575.22. The charity organizations in the cities of Scranton and Wilkesbarre coöperate with the poor directors. The agents of these societies carefully investigate the cases which come before the board and report to the secretary. By the agency of these volunteer organizations fully one third of the applicants for relief are found unworthy, while the other two thirds are intelligently guided in their quest for relief. We have seen in the first section of this chapter that the amount disbursed in poor relief in our territory is excessive. The officers in charge are not wholly to be blamed, for the territory

*A law passed by the last Legislature makes it possible for boards of poor directors to bury the deceased of indigent families if a petition is presented them signed by ten tax-payers.

covered by the average member of the board is large and it is beyond his power to investigate each case, while the remuneration is not adequate to permit him to devote all his time to the work. Of course, the men in charge could do far better than they do, and there is cause for just complaint when men hold the office and are so negligent of their duties as to permit persons to remain on the lists year after year without investigating into their circumstances and also enter others on the roll for no other reason than to enhance their political prospects.

Officials of this character vehemently denounce the interference of the agents of charity organizations and look with suspicion upon the scrutiny of tax-payers. The Tax-payers' Association of Schuylkill county was vigorously opposed by the politicians, and only after a conflict of several years' duration in the court house was it able to acquire such power that the arrogance and impudence of incapable and dishonest politicians were cowed. It will pay the tax-payers everywhere voluntarily to maintain capable agents, free from political influence, to investigate the cases of applications for relief in these coal fields. The public officers are politicians, and the best of them are inclined to take a charitable view of the matter and give help in doubtful cases. One of the best men we knew holding such an office said: "All who come for aid are poor anyhow and should be helped." This is the principle on which much of the out-door relief is given. Of the $26,000 spent in out-door relief in the Middle Coal Field District, over $18,000 is distributed in the Hazleton section. The prodigality with which out-door relief has been distributed in Coal township is criminal. The only conceivable and right way to put a check on this extravagance is for the tax-payers to secure an agent who will revise the list and investigate into the conditions of every claimant for relief. There is no doubt that many now receiving aid can work if a practical test were enforced, and to keep them on the list is to put a premium on pauperism. The unfortunate pauper can only be separated from the idle parasite by an investigation of individual cases by a competent agent of organized charities. The ones now in existence in the coal fields do good work. Others should be organized by the coal operators and thrifty citizens, upon whom fall the burden now due to the extravagance of politicians, who freely scatter that which the thrifty and industrious gather. Patriotism and an intelligent conception of what is good for our society should move our people to action. Pauperism grows under a lax administration, it should be the conscious purpose of society to diminish it. Ransom Poor District has an efficient restriction on out-door relief; the poor directors are annually limited to $5,000 for that purpose. In other districts, however, no such wholesome restraint is exercised and the rule followed by most incumbents of these offices may be said to be that of the railroads: "as much as the traffic will bear."

Fraternal orders also practice charity. When a member is sick beyond the time limit during which benefits are paid, his case is brought before the society and a donation is made. These organizations also profess to take care of the widows and orphans of deceased members, which is sometimes done and sometimes not. The charitable impulse of some of these orders often takes a different form from that of a donation. A committee is appointed which conducts a raffle, when some article, such as a gun, a watch, or even a cow or horse, possessed by the unfortunate, is chanced. Tickets are sold—the widow generally peddles them from door to door. A night is set for the raffle, which generally takes place in a saloon, and the lucky number gets the prize. Sometimes a colliery will aid an unfortunate brother by taking up a collection, when each employee gives a certain sum which is deducted from his pay in the office.

All these efforts, however, are sporadic and the sum collected by these means is given in bulk to the recipient. With few exceptions, the parties aided are thriftless and improvident and the amount collected is soon spent extravagantly. We knew a family so aided spending $8 on two pictures peddled by an agent, and within six weeks after, was again in want. Many of these people have no economic foresight; they live to gratify present desire and the aid given them in ready money is liable to all the abuses of spasmodic efforts to relieve want and results in an immediate profusion of good things soon followed by dire distress. They live in the savage state surrounded by civilization.

It is a part of the Christian faith that contributions be taken for the poor, but the local church lives up to its creed according as it feels financial stress or not. As above stated, the Protestant churches of mining towns are for the greater part weak and all they can do is to meet current expenses and contribute something annually to the missionary enterprises of the denomination to which each belongs. This economic pressure in the life of many churches, together with the absence of want, prevent the establishment of poor funds in many religious organizations. There is in some churches, however, a poor fund to which the members have an opportunity to contribute every time the eucharist is celebrated. Churches of from 200 to 500 members disburse each year from $10 to $50 in charity. In conversation with pastors on this subject they said that very little urgent need is met with among the members of their church and congregation.

In our cities many of the Protestant churches display considerable activity in charitable work and much money is contributed by them for this purpose. Some of them have a home in a quiet country village where they send destitute persons for a few weeks' vacation in summer time. Others have a cottage at their disposal, where young lady clerks of limited means can go for a two weeks' vacation at very moderate expense. Protestant churches also in our territory hold union services on Thanksgiving Day, when an offering is taken up for the hospitals. This collection generally amounts to from $8 to $10 in mining towns. Churches in Scranton and Wilkesbarre contribute annually to the Home of the Friendless, while others gather the soiled and discarded garments of the rich saints to distribute among the poor. Rich churches sometimes also send donations of old hymnals and an organ to a missionary chapel in a mining town, which implements, having sacred metropolitan associations, may yet be used to aid the devotions of the humble, while better instruments are purchased by the city church.

Some of these churches in our towns have other ways to aid

the indigent. When a widow struggles for her children, some charitable persons come together and decide to make a "pound party." A night is chosen and each woman brings her contribution. They go to the home of the widow, present the gifts and spend the evening in hilarity. The good cheer these people bring does as much good to the struggling widow as the gifts. We have also seen Christian associations appealing to the gambling instinct to help the poor, but the most frequent method pursued is to make a social or a concert for the benefit of the indigent.

The Roman Catholic churches contribute to the hospitals, while they also have institutions for orphans and foundlings. The women connected with their various religious orders visit the sick and aid those in distress. The priest in charge provides means for the relief of the needy in his parish, while organizations of the laity, both male and female, practice the virtue of charity among the indigent of the parish.

The liberality of the poor boards in mining communities is such that comparatively small is the number of those dependent on the charity of the churches. These organizations supplement the system of poor relief, but it is questionable whether the sporadic effort they make does not more evil than good. The work of charity has been placed on scientific principles and the sooner all charitably disposed volunteer organizations coöperate along lines laid down by leaders in this work, the better for society. Poverty is a disease which can only be stamped out by the adoption of scientific methods and principles. This should be the aim of all intelligent members of society, and the anthracite coal fields are sadly in need of systematized and uniform methods in the work of relieving the poor.

Individual Givers.

Possibly the most liberal body of men in the anthracite coal fields consists of the politicians. Many of them hold office for two and three terms, draw salaries of from $3,000 to $6,000 a year, and, at the close of their official career, have nothing. Their salaries are far in excess of their capacity, and their social worth would command no such stipend in industrial and commercial circles, but these men have one virtue — they give freely of their wages toward charitable causes. As public servants they are constantly besieged by persons either in distress or soliciting aid in behalf of indigent neighbors. Some of the politicians are liberal from principle ; others from policy. One of the most successful men in political circles in one of our counties passed for years as a generous and liberal man, but when his political prospects ceased by the political law of rotation in office, he pulled the strings of his pocket-book tight and soon became noted as a close-fisted fellow. But taking the politicians of these communities as a body, they disburse more money in charity than any other body of men of equal number.

There are some persons among coal operators who delight to relieve the needy and do much of it. We have had occasion to mention the Hazleton mountains as a section where abuses prevail. There also the higher virtues of charity and mercy are more liberally exercised than in any other section of these coal fields. The ladies in the families of the Coxes, Markles, Wentz, etc., have exercised liberal charity for years among the employees of the above companies, and in addition to their personal services they hire trained nurses to visit the homes of mine employees where some one of the family is sick. These nurses suggest to the mothers methods of treatment and conditions of hygiene which greatly aid the unfortunate. Many families on these mountains have received favors from the ladies above mentioned which they will never forget. We have known wives of individual coal operators elsewhere, who have made unto themselves a name which will live until the last recipient of their kindness has passed away. Many of the individual operators also are generous and largely contribute to the indigent in an unostentatious way. A gentleman who had an idea that the coal operators in the anthracite coal fields were the embodiment of selfishness, came in contact with many of them at the time the Coal Strike Commission held its sessions in Scranton, and his impression was, "why many of these men are generous and kind hearted." It is true. Two kindergartens were maintained in the Northern coal fields by coal operators; many churches have received donations from them; Mr. Markle has built two rooms for the convenience of the young men of Jeddo and Ebervale; the Coxes have given a room for the use of the young people of Janesville; a library is maintained by the contributions of operators in Minersville; the Pennsylvania Railroad Company keeps a man in Y. M. C. A. work in the Lykens valley; the officers of the Delaware and Hudson have one doing similar work in the Lackawanna and Wyoming valleys; and the Anthracite Committee of the Y. M. C. A. draws a large part of its finances from men directly or indirectly interested in anthracite mining. The Delaware, Lackawanna and Western Company gives a building, heats it and keeps it in repair for missionary purposes in a suburb of Scranton free of charge, while many coal companies have freely given valuable lots, upon which churches are built, for religious purposes. The Mechanical Institute of Freeland is a monument to the memory of the Coxes of Drifton. In Hazel township, in 1902, small-pox cases occurred and the township incurred an indebtedness of $1,300 in sanitary precautions. The money was about to be taken from the school funds when the coal companies came forward and paid the bill. Whenever a public movement goes on in the coal fields, the few operators who still live here are liberal contributors to it, while there is hardly a religious organization in our territory which has not on special occasions been the recipient of favors from coal companies.

Pastors and priests are frequently called upon for contributions and many of them give beyond their means to aid those in distress. The anthracite industry has in the past years known periods of depression which inflict great hardship upon the people. An open winter means half time at the collieries and often have pastors and priests known of distress in the homes of the people which they were unable to relieve. Wages are better now than they were in the eighties and nineties, but notwithstanding the fact of the recent advances we are not sure that men will be safeguarded against want, and the spiritual leaders of these people will not lack occasion to exercise the virtue of charity.

Individuals of other classes — professional, commercial, etc., contribute to those in need. During the strikes of 1900 and 1902 the business men, property owners and physicians exercised great charity. Many of these men knew as they gave goods or services that they would never be paid and still they withheld not their hands. After the strike of 1902 was over, many property owners cancelled the debt the tenants owed and said, "We'll start anew."

In the course of our study of the anthracite coal fields, we have found many short-comings in the mine employees. They have many excellencies also and among them is a generous sentiment and a readiness to give in case of need. Their calling is hazardous and in it they are brave and courageous, but when a comrade falls, the tenderness of a miner's heart is not excelled by that of any other class of men, and when the cry of distress comes these men are ready with their dollar to aid. Neither the mine employers nor employees discuss the utility of individual gifts or the merits of promiscuous giving. They give, that is all. The men to whom they give are generally known to them, but if they are not they give to those who are incapacitated through accident. Many of these unfortunates are seen near the pay-cars, and although the miners know that the money they give will soon be spent in drink, they still give. It is done because of sympathy with the unfortunate with whom they are willing to share the wages they earn by hard toil.

The number of paupers in the anthracite coal fields is not large, and in the treatment of those who, either because of accident or adverse circumstances, are in need, what we need is enlightened philanthropy. A philanthropy that will teach and act upon the principle that society is the loser when its members are defective, dependent and delinquent, and that the welfare of the whole is promoted when the good of its needy members is secured. This is the solution of the social questions which press upon us and, without this being kept in view, our philanthropy will do more harm than good. Sentiment has a part to play in this great work, but cold calculation and the adoption of efficient and effective means have a far greater part in the work. To alleviate present suffering and show compassion to the unfortunate is praiseworthy, but we must go deeper and try to create conditions under which suffering, misery and pauperism will be reduced to a minimum.* Biologists tell us that a normal condition is brought about in each organism by a power possessed by each cell to choose the material for its use, but in addition to this there is something like a common volition in the mass of cells making up the whole. In society each individual should be kept alone to choose that which his intelligence suggests is best for his use, but there is also such a thing as the common volition of the group which consciously seeks the well being of the whole. To guide this conscious purpose of society is the interest of all and it can be done only by an enlightened philanthropy which sacrifices temporary enjoyment for greater future benefits. Above the selfish interests of the few stands the welfare of the masses, and the army of paupers, defectives and criminals will only diminish as society sees that each child possesses equal advantages for gaining a livelihood and contributing to the welfare of the whole.

*James E. Roderick, Chief of the Bureau of Mines, in his report for the current year, asks that provision be made in some way for those who are incapacitated from working in and around the mines. He proposes a tax of half a cent a ton on all coal produced, which he estimates will amount to $375,000 annually, to be evenly divided between the owners of coal land and operators. Some such scheme should be instituted to provide for the maimed, widowed and orphaned of these coal fields, but before this is done the whole question of compensation for accidents in the mining industry ought to be thoroughly investigated by a competent Commission and the work of legislation placed on scientific principles. If otherwise done we will only have greater wastefulness than now exists in the relief of indigent mine employees, and pauperism will increase.

CHAPTER XII.

POLITICS IN MINING COMMUNITIES.

1. The Men who Run for County Offices. 2. The Men who Sit in Council Chambers. 3. How are our Boroughs Managed? 4. The Board of Health in Mining Towns. 5. Mining Municipalities Owning Public Utilities.

The Men who Run for County Offices.

The counties of Lackawanna, Luzerne and Schuylkill are largely governed by the vote of the mine employees. The official vote cast in these counties in the fall of 1901 was 87,648, of which 56,258, or 64.18 per cent., was cast by electors employed in and around the mines.

The character of the candidates for office in these counties is very different from that required in adjoining rural counties. A man who will not liberally contribute funds to keep the "boys" in good humor can not be elected, while such conduct in a candidate for public office in an agricultural community would result in his defeat. The government of our counties is just as good as is the conscience of the average elector. The quality of men chosen to fill positions of public trust will not rise higher than the level of the public conscience, and before greater honesty and integrity can be attained in our administrators, these virtues must rise much higher than at present in the electors of our counties.

All the men connected with the political machines in our communities are not corrupt. We have personally known some of them who are conscientious, honest, and patriotic, and none knows the abuses which prevail in our political life better than they. These misdeeds they heartily deplore, and many of them have gone so far as to sever their connection with the organization and denounced the knavery and fraud which have alienated them.

Some of Our Future Citizens.

In every election the machine is active and, notwithstanding the various devices inaugurated to weaken its hold upon the process of nominating and electing candidates, it triumphs over all schemes to defeat its purpose, succeeds in bringing its men before the public as regular nominees of the party, and elects them. What chance has the ordinary citizen to fight these men who devote all their time to "fixing things" their own way and who command a floating vote that will, under ordinary circumstances, always decide the election? It is the power of the machine in our counties that makes reform so difficult. Men guilty of misdemeanor in office are not punished, and a band of reformers who bring suit finds it next to an impossibility to have opinions rendered in the cases tried. A notorious politician, when tried in one of our courts, was asked if he had done much "dirty work" in politics; he said "yes." "For whom?" "Oh, for many, and among others for his honor on the bench." The judge bowed his head in shame. Men, who in social and religious life occupy places of honor, when dominated by political aspirations stoop to despicable ways to secure votes. One of our Sunday-school superintendents aspired for a seat in the Legislature and, while earnestly engaged in sacred things, his money furnished beer to a crowd of ward-heelers in another part of the city. Few men, aspiring for public honor, have the grace to resist the venality practiced by politicians. A prominent official of the Delaware, Lackawanna and Western Railroad desired office and asked the men in control of the machine to launch him; "Certainly," was the reply of the politicians, "and first of all we want $200 to be distributed in the saloons." The man was innocent of the ways of politicians and immediately withdrew his name. And what wrecks are recorded in the political world in these coal fields! Some of the most successful politicians in Lackawanna, Luzerne and Schuylkill counties were, before they entered politics, honorable and useful men, trusted and respected of all. To-day, those who still survive rank with the most corrupt and tricky in the State.

A successful Congressman from the anthracite coal fields a decade ago spent from $5,000 to $6,000 during the campaign, now from $25,000 to $30,000 is spent, and in a vigorous contest in the fall of 1902 one of these candidates is said to have spent twice that amount. Conventions, where nominations are made, have become in our counties bargain counters where candidates openly buy votes, and the delegates boastfully exhibit the money they secure by selling them. Sums from $75 to $200 were recently paid and the men receiving the money attempted not to conceal the fact. An ex-boss in politics in one of our counties in a recent public address protested against bribery, corruption and commercialism in politics and said that the electors were sold "as sheep in the shambles." A man in the crowd said: "He ought to know; a few years ago he brought into the county $10,000 to buy up delegates for the state machine."

All this is in striking contrast with the condition of affairs in these regions forty years ago. Then it was very difficult to get men to attend the county conventions, now the number of candidates anxious to go is large. The corrupt use of money has done its deadly work and the electors have degenerated. Venality has permeated the mass of voters and all nationalities are equally tainted. Men deliberately offer their votes to the highest bidder, while cases become more and more frequent of men taking money from both parties and voting as they will. A staunch Republican sold his vote for $2 in a recent election, and when his friend remonstrated, he said: "I'll sell my vote to the d——l for $10."

Representatives of the state machine are sent to these counties to debauch the electorate and train its members in knavery. In the fall of 1901 they came to Schuylkill county to buy up delegates appointed at the primaries to attend a local convention, where two delegates were to be chosen for the State convention. The agent offered an anti-machine delegate $20 to stay away from the convention, and to another man $25 to go out of town. Two delegates, who were sent to vote against the machine, were bought over by the agent for $70 each and two railroad tickets to the Pan-American Exposition.

The increased expense incurred in running for county offices has resulted in monopolizing them. Men of limited means cannot hope to secure nominations. A candidate for judge of common pleas recently asked the county machine to endorse him. The committee was willing on condition that he would send them forthwith his check for $1,000. He refused and denounced the exorbitant demand as un-American and ruinous in its tendency to our free institutions. He was none the better. The machine went on, placed a man in the field and elected him.

Monetary considerations are not the only motives which drive men into politics. It is really gambling and the gambling instinct has here free play. Many men, who are strictly honest and upright in business, will cheat, bribe, prevaricate and defraud in politics. One of the most successful politicians in Schuylkill county was a born gambler. His opponent in one of his many political contests happened to be an intimate friend with whom he transacted much private business. One day, the gambler came to his friend, whom he knew to be a man of sterling integrity and of high repute among the electors, and said : " John, in all our business transactions I have dealt justly by you, but in this fight I am going to lick you and that by fraud. I couldn't be honest with my own mother in a political contest." He kept his word, but soon after election he found things too tame. He left for the turf and his friend filled the office for the term, which was his by right if the votes of the electors had been honestly counted. Monte Carlo has few more thrilling tales to tell of the evils consequent on gambling than has the politics of our counties. One of the most recent is that of a young man in the Northern coal field. He had labored hard to save a few thousand dollars and had built two houses. He was popular and thought he could get a county office. He got the nomination and was assured by his friends that his election was certain. The contest developed into a very hard one and after the few thousand dollars he had in the bank were spent, he mortgaged his houses and staked all the savings of his life on the fight. He was defeated. The disappointment was more than he could endure and in six months he died of a broken heart. We knew a man in medium circumstances staking all he had on an election and obliged to borrow money to get his dinner and pay his fare home the day of election. It is running for a prize; if the candidate wins he is saved, if he loses he is ruined. An officer of Luzerne county confessed that he had spent in a contest $3,500 in cash and lost three months' time going the round of the 300 precincts. The office was worth $4,000 a year. He was now running for the second term and had about $7,000 saved. "If," he said, "I lose, I will be worse off than when first I sought the office." In the election of last November one of the candidates in Schuylkill county spent over $5,000. He was defeated and now has a mortgage of $3,000 to pay.

The amount of money spent by candidates is not uniform. It depends on the value of the prize sought, the number of men on the ticket, the wealth and the generosity of the applicant for office and his capacity to make a bargain. Most men are willing to spend a sum equal to one year's salary of the office they seek. Some candidates do not know exactly how much they spend, others keep an accurate account of every dollar. The temper of the people also has much to do with the effect of money spent on election. If a wave of indignation moves the masses in one direction, money, lavishingly spent, falls short of the desired result.

The campaign is managed by a county committee which assesses each candidate according to the value of the office which he seeks. This money is spent for printing, revising the registry of voters, engaging speakers to stump the county and manning the precincts on election day. The assessment varies according to the number of candidates running. It costs candidates generally from $30 to $80 to register, and after the nomination they are assessed from $200 to $1,000. They are then supposed to go through the county for about three months and, each day, a candidate will spend from $15 to $20. It will also cost him from $400 to $600 to man the polls on election day. Some 25 or 30 papers will be anxious to serve him, and he must spend on each of these an average of $10. If he is "green" the newspapers may run a heavy bill. A rich candidate appeared in Luzerne county not long ago, who was not versed in the way of politics. One of the papers asked him if it should boom his candidacy and the man said, "Certainly, go ahead." At the close of the campaign a bill for $380 was presented him by the firm and the editor sent him a bill for $50 extra for the glamor he threw over the candidate in his glowing articles.

The free use of money before and on election day proves that the venality of electors is great. No sooner has a candidate declared his intentions to run for a nomination than he is besieged by an army of parasites who represent some club, or organization. Heads of political clubs also come, each affirming that he can command a certain number of votes. Most of these clubs are associated with saloons and their quest is "free beer." One of our saloons had on tap "free beer," for three nights of each week, for three months previous to the last election. It was the contribution of the candidates to the political club that watched over their interests in that ward. Fire companies are also after the candidates and these "hose houses" are supplied, before election, with something beside chemicals for fire purposes. On the Fourth of July, 1902, a candidate for the Legislature was besieged by 75 parties, each of whom wanted a keg of beer to drink in his honor on the holiday. Before the day was ended the man had to go to hiding to avoid the plague. Another candidate of last fall was notified by a club 150 strong, that it would hold a supper on a certain date and would be pleased to have him furnish the necessaries for the feast. Clubs of all sorts arrange their annual dances about election time and send a pack of tickets to each of the candidates. The Sclavs have brought their efforts down to business principles. Their clubs send their agents to the candidates to inform them that the organization commands so many votes and is open for bids and that the highest bidder will get them. Even religious organizations are not above availing themselves of an approaching election to wheedle money from candidates upon whom they have no claim whatsoever. The army of men asking legislators for passes on the railroads is innumerable.

One of the legislators lately kept record of the number of applicants and his list at the close of one month contained 250 names. He served notice on his friends, saying "The railroads did not come to my possession with the office." The Sclavs follow the cue given them by the Anglo-Saxons, but the latter have very little more to show these apt pupils of political corruption. In many of our towns, and especially in Schuylkill county, the Sclavs hold the balance of power.

The price paid for votes depends upon the contest. If it is very close the price may advance to $2 and $3 a vote. Generally, however, the rate does not exceed 50 cents each for Anglo-Saxon votes, while those of the Sclavs come cheaper. A Sclav political boss, controlling 100 votes in a mining camp in Luzerne county, offered them to one of the agents of candidates for office for $20. The manager said, "Too dear, we can get that kind of votes for ten cents each." It is impossible to say what percentage of the electors are venal. In every town in these coal fields there is a gang of "floaters," strong enough numerically to decide the election. These persons mercilessly bleed the candidates for office. When the election is uncertain, money is freely used on election day to buy these "floaters," and men who know the power of money say as did Louis XIV. on one occasion: "After all, it is the last louis d'or which must win."

Is it strange then that these men, who have been bled by an army of electors before they were elected to office, are anxious, when in office, to replenish their treasury as rapidly as possible and, in their anxiety, forget to do so in constitutional ways? Few of these men are above corruption and the standard set by the famous Pitt, who said: "I will not go to Court if I may not bring the Constitution with me," is seldom thought of by the best of them. They are the victims of despicable schemes of electors, who organize for no other purpose than to hold up the candidates, and when elected to office their opportunity comes and they are anxious to reap as much as "there's in it." This is one of the chief reasons that they plan jobs and give them to favorites, that they spend the people's money in clerk hire and needless printing, and that each department, with the connivance of the commissioners, largely follows its own will in its expenditures. One of the scandals recently exposed in the Pottsville Court House in connection with the Schuylkill Almshouse was, that the poor directors, having the power to appoint officials, elected those only who consented to a "rake-off" of a certain percentage of their salaries. The only restraint upon the politician's greed is volunteer organizations which are comprised of competent men versed in the law and the Constitution of the State, so that they may prosecute those who take money while in office in other ways than the legal and constitutional ones. The Tax-payers' Association of Schuylkill county has done good service in this regard, but it has only been able to effect its purpose by continued vigilance from year to year, and persistent appeal to Court to compel the officers to abide by the Constitution and to force controllers to refund moneys paid contrary to the law. But notwithstanding the vigilance of these public-spirited men, the politicians keep them busy. In many of our boroughs tax collectors are appointed whose bonds are worthless and the large balances due the county cannot be collected. Items of needless expenditures appear in the Controller's reports every year and the appeals made in former years show that "there is little probability of any action on such matters." Last year the treasurer refused to turn over $2,500 interest money into the county treasury on the plea that he did not know who was entitled to the cash. The Coroner of Schuylkill county investigated deaths which needed no inquest, held irregular inquests, returned duplicate charges, etc., which if withheld, would have cut down that year's earnings of about $10,000 fully 50 per cent., but notwithstanding the protest of the association he was paid in full. Clerks and deputies were paid salaries when the earnings of the offices were not sufficient to meet them. The assessors' fees have increased each year, so that now it amounts to $20,000, one third of which ought to be cut off. The jurors drawn in all our courts have been under political influence, and in each county the costs of ignored bills have been placed upon the tax-payers, which accounts for the large number of needless suits returned by justices of the peace. The above-mentioned association has put a stop to this practice in Schuylkill county by placing the costs on the litigants, which has had the salutary effect of reducing the number of suits one half.

The members of the Legislature are also said to be anxious to replenish themselves. The following anecdote well illustrates the attitude of the average member. One of the men elected to the Legislature of 1901 was congratulated on his successful resistance of the bribes offered by the machine which, in some instances, had gone up to $1,500 and $2,000 per vote, when he replied: "Yes, but I wish to tell you right here, that they have come damned near my mark."

Ex-Lieutenant-Governor Watres, speaking last summer in Altoona, criticised the Legislature of 1901, which was dominated by the party to which he belongs, in the following terms: "To-day the proud State of Pennsylvania stands before her sister States with shame and confusion of face. The Legislature of 1901 brought merited reproach, not only upon the party, but upon the Commonwealth. Commercial politics have well-nigh wrought the party's ruin. Bribery has run rampant. Election frauds have been winked at. The people's rights have been laughed at, and those who plead for them have been turned to scorn."

The Men Who Sit in "Council Chambers."

The character of our boroughs and townships is not uniform, but the character of the people reflects itself in the personnel of the council and board of directors. In Schuylkill county there are two boroughs not far from each other; the one is conservative and economical in its management of affairs; the other has always been notorious for its wastefulness and recklessness in the management of public funds. In Nanticoke, Luzerne county, the character of the town council is unsavory and many respectable men will not consent to run for a seat on it, but the board of directors is above suspicion and the best citizens of the borough consider it an honor to be elected to the body. In Lackawanna county there are two boroughs whose affairs are managed very differently. The one council voted a street railway franchise into the pocket of its president; the other body insisted on taxing the property of the street railroad in the borough and turned several hundreds of dollars annually into the public treasury. One borough spent $1,200 on a fire alarm system that constantly got out of order, while another got satisfactory service out of a $400 one. The borough of Nanticoke, with 12,000 population, spent about $8,000 on a municipal building, but Edwardsville, with only 5,000 population, spent $18,000 on such a building, $8,000 of which was for extras. These differences in the management of boroughs and the use made of public funds depend upon the personnel of the council, which again reflects the character of the electors in the borough.

The moral quality of the council or the school board of directors seldom rises to the level of that of the average citizen in the borough. We have known men of unquestionable integrity positively to refuse the request of their fellow-citizens to enter the council of a borough, because they would not be seen in company with the men who occupied seats therein. A man of honest purpose cannot long occupy a seat among selfish and unscrupulous men, bent on serving their own and not the people's interest. He is helpless to effect reform and in disgust he leaves a body of men the majority of whom is dominated by egoism and rudeness. Reformers, who have tried to prevent dickering and jobbery, have been subjected to such abuse and scurrilousness by foul-mouthed scoundrels as is only heard in quarrels in low bar-rooms or in fights among the outcast of society. Few men are willing to submit to this and accomplish nothing for the public weal. The character of town councils in many boroughs in the counties of Lackawanna and Luzerne has been such that the Lehigh Valley and the Delaware & Hudson companies have forbidden their foremen and assistant foremen from entering these bodies.

The Irish-Americans are among the most successful in the political life of these coal fields. In a borough where they formed only 35 per cent. of the population, 74.8 per cent. of the total number of councilmen for the last ten years was of Irish descent, and out of one hundred persons who held remunerative offices in the borough during the decade, 81 per cent. was Irish. This aggressive people is far more successful in getting into office by the vote of the electors than the representatives of any other nation in our territory.

The saloon is also a very important factor in our politics. All through our boroughs and townships the primaries and elections are held in saloons. There are no other places available, for they cannot be held in school buildings, and churches will not open their doors. All the ballots cast in mining towns are polled in the stench of beer, and although saloons are legally closed on election days, men connected with the election get "tight" before the day is over.

Saloonists also are directly interested in elections. In the borough of Olyphant for the last decade (1891–1901) the saloonists formed on an average one third of the number of councilmen. Besides these, others were indirectly connected with saloons, so that directly or indirectly 43.3 per cent. of the seats was under the control of the saloon. In nearly all our boroughs 2 or 3 out of 12 members in council are saloonists. These men carry far greater influence on the action of the council than their numerical strength would suggest. The saloonkeeper is generally an active man and manages to organize a clique which regularly meets in his house. This coterie, held in line by drink and cigars, invariably does the bidding of the chief. In Nanticoke, last year, a saloonist and a band of Polish councilmen dominated the affairs of the borough. It is the rule in most boroughs that the money received from licenses is to be spent in road repairs. The saloonists generally get their men into this work and the borough orders find their way into their business places. A man who had closely watched the manipulations of saloonists in selecting men for street work said that from 40 to 60 per cent. of the orders paid these men went into the hands of the saloonists. When the treasury of the borough is low, some of these men cash the orders at a 25 per cent. discount.

The Sclavs enter politics more and more each year. A Polander was recently mayor of Shenandoah and made as good an official as the average Anglo-Saxon incumbent. A Sclav justice of the peace in Nanticoke was the first to comply with the law to return all fines collected under the compulsory school attendance act to the school funds. In 1891 none of the council (12 in number) of Shenandoah was Sclav; in 1902 five of the members were. In this largest borough in Schuylkill county they hold the majority vote in three out of five wards, and 46 per cent. of the electoral vote is in their hands. In Nanticoke the majority of the council of 1901 was Sclav and in 1902 five of the members were of that race. In addition to this they hold two seats on the Board of School Directors; one of the two justices of the peace is a Sclav; so also are the tax collector, the borough treasurer and the street commissioner. The Sclavs are shrewd. If two candidates from among them try for the same office, a caucus is held, a vote taken and the one who gets the largest number of votes runs. The Sclavs seldom divide their vote; they invariably move in a mass. Every year they enter politics in larger numbers and the very men who, a few years ago, abused and maligned the Sclav as the curse of the anthracite coal fields, now refer to them as "American citizens, loyal to the American flag and proud of the land of their adoption; they are striving for the elevation of themselves and their children and endeavoring to merit the esteem of the people with whom they are associated." Politicians are keen to read the signs of the times. Sclavs, who have entered the councils of some of our boroughs, are not so venal as the average Anglo-Saxon member. But, somehow, these people are quick to learn the tricks of their predecessors and the danger lies in the Sclavs imitating the "spoils system," which is now the basic principle of our political life, and falling into deeper corruption than is now practiced in our boroughs. As long as the elections are dominated by corrupt and unscrupulous men, the type of borough officials will not improve. A radical change in state, city and borough politicians must come before men elected to office will feel it their duty to conduct the affairs of the borough as they would those of their own.

The history of some cases, tried in our courts, of public officials abusing their power is the best evidence of the need of reform. A coterie of men in the councils of Scranton formed a combination to favor the street railway company for the consideration of being entered on its monthly pay-roll. The Municipal League of Scranton exposed corrupt practices of public officials in the city government which stirred every honest citizen to indignation, and yet the public censure was not strong enough to demand that these men be punished to the extreme limit of the law. Intelligent men argued, they are no worse than others in office, and corrupt practices have grown by years of indifference on the part of the people to dishonest administration of municipal affairs, and why should these men be punished above their fellows equally deep in crime? The public conscience also demanded the prosecution of the companies which gave the bribes. This was not done and the reason, as some alleged, was that "some high heads would be made low." The cases of Olyphant, Winton, Lackawanna township, the almshouse of Schuylkill Haven, etc., all reveal a sad condition of affairs, and when the culprits are brought before the bar, all that is required of them is that they resign from office and restore the money which they have misappropriated.

In the quest for public office in boroughs and townships money plays an important part, though not so great a part as in county offices. The councilmen spend from $25 to $150 in running for office. This is spent among the "boys" to secure them cigars and beer. The average member is anxious to get something out of the borough office, although he knows before he is elected that there is no remuneration in it. They generally ask, however, "what's in it," and, before their term is over, they have found out ways whereby their beer and cigars have been secured and, in some instances, something more. There are in each borough several remunerative offices at the disposal of the council, and when these are distributed the shrewd politician invariably remembers some relative or friend who is just fitted for the position.

How Are Our Boroughs Managed?

In the hands of men of the character we have described, it is not strange that municipal government is corrupt. In the three counties Lackawanna, Luzerne and Schuylkill, over two and a quarter millions of dollars are annually handled by the men in office, which amounts to about $2.79 per capita of population. Besides this, there passes through the hands of borough and township officials taxes to the average amount of $1.70 per capita of population, which aggregates to about $680,000 in mining towns. We have left out of the count the appropriations made for public schools as well as the State appropriation received in each borough and township. These large sums of money annually distributed is the lure for the politicians, and the service rendered for these vast sums is often more mischievous than beneficial to the municipality. The sums collected, however, are not enough for the politicians; in every borough there is a load of indebtedness which in many instances has passed beyond the legal bound set for the borrowing capacity of the boroughs. In Olyphant and Winton boroughs and in Lackawanna township no one knew the indebtedness incurred until the court appointed auditors to find out and then decreed a special assessment to liquidate it. In the borough of Olyphant the treasurer had been allowed for years five per cent. on all money received and three per cent. on all money paid out, until the court, moved by the Delaware and Hudson Company, ruled that the treasurer received compensation far in excess of the services rendered. In this borough also councilmen drew on the treasury in one year sums aggregating $629 by taking trips to adjoining boroughs in the alleged interest of the people, for which they charged $2.50 a day, which was also declared illegal by the court. In Lackawanna township the auditor appointed by court spent eighteen months wading through the intricacies of the accounts left by incompetent and careless officials. His investigation resulted in marshalling an indebtedness of over $41,000, and his own bill of expenses in the work amounted to over $5,000. In the borough of Olyphant the auditor found the indebtedness amounting to over $42,000. In Edwardsville the indebtedness of the borough has passed its limits and in order to borrow $6,000 to construct a sewer it had to obtain special permission from the court. In Nanticoke the debt has reached the limit of the borough's borrowing power. One borough of 6,000 population gives its treasurer about $1,000 a year, in another of double the population the salary of that official is only $250, and a storekeeper in the latter place offered to take the office without compensation; he would get his returns in filling borough orders over the counter.

The average politician feels no compunction in taking money from the borough or county treasury without giving adequate service in return. In all our counties the salaries are out of proportion to the services rendered, and the men drawing these salaries could not make half so much if engaged in any other line of business. One of our sheriffs makes nearly $100,000 in three years, and a district attorney over $10,000 a year. Clerks in court houses draw from $75 to $100 a month when they could not earn more than half that amount in clerical work open to competition. In February last, the Tax-payers' Association of Schuylkill county moved court to serve an injunction on the controller forbidding payment of the salaries of clerks in the court house, because their writing was so poor that the records could not be read.

In boroughs, there is a woeful lack of common honesty in the performance of labor for the municipality. In one of the best managed in our territory the president of the council said that labor of all kinds costs the borough from 25 to 35 per cent. more than a private individual could get it for. And this lack of conscience in the discharge of public service characterizes men whose education and training ought to raise them above it. A lawyer, in one of our boroughs, asked for $50 to meet the incidental expenses of a suit. He got the cash and at the close of the year turned in the bills for the incidental expenses incurred. The council of one of our towns wanted a borough map. An engineer was engaged at $7 a day and after his bill amounted to $977.85 he was discharged. A member of the council asked, after the bill was paid, "Where is the map?" No one knew. About a year later, a servant in one of the saloons was cleaning the cellar and had collected the refuse to burn it. Some one happened to see a roll amid the mass, and lo! it was the missing map which cost the borough nearly a $1,000! Two streets were surveyed in the same borough and the work cost about $650. In another borough, two constables, hired to keep the peace, were found one evening drunk, lying in the gutter, and although some of the indignant citizens demanded their removal the indulgent hand of the machine protected its minions.

Of all the inefficient services rendered the boroughs possibly the worst of all is that of road repairing. A prominent citizen of Edwardsville said: "Last year [1902] $2,000 were spent on the roads of our borough and I would like to know where is the improvement?" Another citizen of Nanticoke said: "$3,800 was spent on our roads and not $500 worth of work given, and for the year 1902–1903 the street committee asked for $4,500 and the park committee for $1,500 to paint, etc., a small building erected in the park." In the borough of Olyphant a sum has been spent on the streets during the last decade which, if economically used, would have been ample to give brick pavement on the principal thoroughfares of the town. Men are paid $1.50 a day for eight hours' labor on the streets when they do not do 75 cents' worth of work.

And not only do petty officials plunder the public treasury but leading citizens in boroughs also shirk their part of the burden of taxation if by any crook they can do so. In a borough where a sewer was put in, three or four professional men, because of their superior knowledge of the law, escaped paying their assessments, while the mine employees, who owned property, paid their share. In another borough where improvements in the street were made, about a dozen leading business men refused to pay their part because of a technicality whereby they escaped their obligations. The debt, which amounted to nearly $1,000 and had been standing for several years, was removed from the assets of the borough by a motion from one of the councilmen who said "It's a dead dog anyhow." In each of the above instances the service was rendered and the men received the benefit of the improvement, but their sense of moral obligation was such that they shirked their duty when they knew that they could not be compelled to pay the bill. It was as much of a robbery of public funds as is the plunder of petty politicians. Men who serve boroughs believe they can render less service for the money received than when performing work for private parties. Coal haulers were detected, a few years ago in two of our boroughs, delivering coal, which should go to the public schools, into the cellar of a parasite and nothing was done about it. Another man had pocketed the rent of a borough property for years and when discovered he had the audacity to deny all and the politicians believed him. One of the presidents of our councils was known to serve a company which asked favors, and in order to clear himself of suspicion he asked his clique to pass a vote of confidence in him, and so it did.

In officials as well as in the leading men of these boroughs there is a lamentable deficiency in moral sensitiveness regarding public funds, and although they demand business integrity and commercial honesty in private affairs, they connive at dishonesty in public servants and are themselves not above shirking their obligations to the public when they can safely do so.

Possibly a large part of this callous indifference to the public weal and neglect of public duty is due to the fact that coal corporations pay a large percentage of the taxes. In Lackawanna county the corporations pay one third of all the taxes. In Schuylkill county, the Philadelphia and Reading company alone pays annually over $80,000 taxes, while all the anthracite coal companies pay of State taxes the sum of $429,949.69 and the anthracite railroads $837,757.32. In thirteen townships, where coal is mined in Schuylkill county, the coal companies pay from 84 to 97 per cent. of the taxes, the average being 91.6 per cent.; the total taxation amounts to $64,573.03. In three townships, where coal is mined in Northumberland county, the coal companies pay 72 per cent., 88.5 per cent. and 60.4 per cent. respectively, while in Conyngham township, in Columbia county, they pay 77.6 per cent. of the taxes. In the borough of Lansford, the Lehigh Coal and Navigation company pays 60.9 per cent. of the taxes. Coincident with this collection of taxes from the coal companies in the above mentioned townships is also the exoneration of a large number of individual taxpayers in these communities. In the thirteen townships in Schuylkill county the total taxation which fell upon the persons residing there amounted to not over $6,000, and $3,697 of it was exonerated. And in the three townships of Northumberland county the people are supposed to pay $7,701 for school purposes, while the list of exonerations amounts to $3,002.

From these facts it is evident that civic righteousness is not developed in the average citizen of these territories. Fifty per cent. of these men has no scruple in taking municipal privileges without paying its just share of what they cost. Politicians are as reckless in spending moneys derived from corporations as they are ready to exonerate their "friends" from their obligations. In this study one cannot avoid the conclusion that the development of the individual conscience to a right appreciation of civic virtue and morality must precede all reform in municipal government in these coal fields.

The "Board of Health" in Mining Towns.

In no sphere is the greed of politicians and their neglect of the people's interests more conspicuous than in the inadequate attention given to the sanitary conditions of the boroughs in the anthracite coal fields. No department of public service is as directly related to the health and comfort of the people, and yet in a large percentage of our boroughs little or no attention is given the subject. It has been proved in thousands of instances that proper drainage, a plentiful water supply, the removal of decomposing organic matter, improved paving, scavenging, and public cleanliness, reduce the death rate, and that many of the scourges which prevail in our villages and towns arise because of lack of attention to these things. The politicians care little for this; in most instances the funds they handle are too small to go the round, so that little can be spared and with reluctance appropriations are made for the purpose of promoting the public health.

Barbarous peoples, who know nothing of the principles of hygiene, can be excused when poisonous matter accumulates around their dwellings; but a civilized people, in whose schools are daily taught the principles of hygiene, ought to guard against nuisances which menace the health of the people. Yet there is not a borough of 4,000 or more population in our territory where foul and fetid spots which breed disease and death are not found.

In boroughs where boards of health are organized, the members are generally minions of the machine. When a board was appointed in one of our boroughs, the first question asked by one of the three members was, "What's in it?" In another borough of 12,000 population we asked the health officer, "What was your death rate last year?" He said, "I couldn't tell, I think the certificates are in the office." No record of deaths was kept in this town for years and yet it had been visited by small-pox, while diphtheria, typhoid and measles had carried away scores of the inhabitants. Why should politicians care for the health of the people and the diseases which kill? There is nothing in that for them. Registering voters and preparing slates pay, but the statistics of life and health of the people, what is there in that?

The water supply of many of these mountain towns is bad. In one section of the coal fields the people complained that they had to drink water contaminated with culm. Typhoid fever prevailed and the secretary of the board of health wrote: "Whether or not this [mixing of the culm with the water] contributes to our increase of typhoid fever has yet to be settled." In 1900, six towns in the Wyoming and Lackawanna Valleys sent samples of the water used by the people to be tested: two were labeled "suspicious," two "usable" and two "impure." This water was used by 100,000 people and the municipal authorities had it in their power to compel the companies to furnish wholesome water to the public. In another town the water supply, in the summer months, ran through a marsh where decaying organic matter accumulated. The State authorities were appealed to, but all they could do was to refer the complainants to the town council, which was indifferent, for they drank no *water*. In Wyoming, there is a marsh which has bred malaria from time immemorable. Attempts have been made to drain it, but, like all else done at public expense, the work was poorly executed, and to improve things several coal companies took the liberty to turn in the refuse water from their breakers and washeries, so that it is to-day a worse plague spot than ever, breeding malaria from which suffer the thousands of Wyoming, Exeter, West Pittston, etc., but the municipal authorities do nothing.

In 1898 and 1900 the State Inspector was called to two boroughs in Carbon county, where nuisances existed which menaced the public health, on three different occasions. He found parts of the towns filthy and unhealthy, while many cases of diphtheria prevailed, arising chiefly from defective or no sewerage. But all the Inspector could do was, in the one case, to recommend the organization of a board of health, and in the other to advise the board of health and the council to work in harmony, so that the streets and gutters might be cleaned, the water closets, out-houses, pig-pens, and stables be disinfected, nuisances be abated and the lives and health of the little ones be protected. A visit to the towns of Mahanoy City and Shenandoah on a summer day will reveal pest holes which contaminate the air, while the foul creek which runs through the center of the former place ever emits poisonous and noxious gases. People live in basements in these towns which would be condemned by the municipal authority, if it insisted on hygienic conditions in the abodes of the poor. People grow accustomed to noxious smells and they do not mind them very much, but when a flood comes and sweeps the contents of a vault into the cellar they protest. It matters little, however, for the poor live in cellars and the proprietor of the building generally has a pull in the ward, so that he is not made to do what a righteous administration should compel him to do.

A recent report of the inspector of public health of Luzerne county is little better than a lamentation, because of the inefficiency of local boards of health and the state of criminal negligence into which they had fallen. He says that the public is culpably indifferent and the officials careless, and that there are no funds to hire inspectors or pay doctors who report contagious diseases. Of the sixty-one cities and boroughs in the anthracite coal fields where boards of health ought to be organized, only 50 per cent. make an attempt to comply with the State law, and the majority of these do their work very unsatisfactorily. In twelve boroughs, in 1900, the number of diphtheria cases amounted to 2 per cent. of the population. This disease is peculiar to children under ten years of age which form about 17.5 per cent. of the population, so that in these towns two cases of this dreadful disease prevailed among every 17.5 of the children, or 11.4 per cent. of the population under ten years of age was so afflicted. In the same year 110 persons died of typhoid fever and 45 of scarlet fever in the above territory. In ten boroughs, with an aggregate population of 88,706, there were 1,818 nuisances abated, or 10.3 per cent. of the houses in this area were so affected in one year.

Our towns and villages are located in the hills, where a plentiful supply of wholesome water could be easily procured and good drainage effected, if engineering skill and considerations of public health dominated. The principles of hygiene are taught in our public schools; they ought also to be taught in council chambers. Healthy children and happy homes would be seen everywhere in these coal fields if municipal administration were in the hands of intelligent and public-spirited men and the lethargy and ignorance of the people were more closely scrutinized and disciplined.

The state law requires that a board of health be organized in every borough. This wise statute is extremely important

A Mining Town. (1,600 feet above sea level.)

when we remember that in every borough throughout the anthracite coal fields are found colonies of Sclavs, most of whom are ignorant of the first principles of cleanliness. The sections of our towns inhabited by them are often reeking with filth, and the noxious smells which arise from the streets and alleys testify to the poisonous gases generated by the uncleanly habits of the people. This accounts for the high death rate among the children of the Sclavs, and the adults of this part of our population are only able to live in these unwholesome surroundings because of a strong physique secured under very different conditions on the farms of their native homes. These people have come to a higher civilization and no better discipline could be administered unto them than the supervision of a competent health officer in every town who would, with uncompromising attitude, insist upon the observance of the laws of hygiene as laid down by the experience of civilized communities. Indeed, much tuition in this respect is also needed by the Anglo-Saxons in these coal fields. The majority of the ills to which our people are subject rise from ignorance of the conditions of health in feeding, clothing and housing the body. In this department of public life, as well as in that of charities, there is little hope of progress as long as the work to be done is left to politicians. Science has its teachings upon these questions and its dictates must be heeded if social progress is to be attained.

MINING MUNICIPALITIES OWNING PUBLIC UTILITIES.

State socialism finds favor in many parts of our country, and, among others, some towns * in the anthracite coal fields believe

* Dr. Frank Julian Warne called attention last fall to the growth of socialistic sentiment in the anthracite regions in a series of interesting articles on "Sclav Invasion of the Anthracite Region," published in the Philadelphia *Ledger*. The following figures give the socialistic vote cast in the counties of Lackawanna, Luzerne and Schuylkill for the years specified:

	Socialist Vote.	Per Cent. of Total Vote Cast.
1900	812	0.9
1901	2,488	3.1
1902	18,406	19.5
1903	2,711	3.3

In Carbon county, where the socialists seem to have a larger following than in any other section of the coal fields, the socialistic vote this year, as compared that municipal ownership of public utilities is the right thing. The politicians seem to favor the idea, and it may be of interest to the advocates of this system briefly to record the experience of some of our municipalities which have experimented in this line.

The borough of Olyphant, Lackawanna Co., has constructed an electric plant which has a very interesting history. The work was commenced in the year 1892. Trouble began in the construction of the foundation wall. The contract was awarded to an unreliable party who, having taken $273 for some work done, abandoned the work and the borough was forced to complete it by hiring a competent mason at $3 a day. The sum of $8,000 was spent on the plant and the air was full of rumors that more had been paid for the machinery than it was worth, and the cost of running the plant in labor and supplies was unreasonably high. Before the electric plant had seen two years' service, the politicians said it must be enlarged by the expenditure of another $15,000. The council advertised for bids and those of the firms which asked the highest prices were accepted, notwithstanding reliable competitors were $2,000 lower. A suit followed and the contracts were annulled. Nothing daunted, the council executed another with the very same firms in the year 1895. The engine and boiler of the old plant, for which the borough had paid three years previous the sum of $4,000, were sold for $800. Operations were begun on the new plant and citizens interested in it complained that the foundation wall was not in accordance with the specifications, but the work went on. The court, however, served an injunction to prevent the council from issuing the bonds for the $15,000 loan. The case was carried to the supreme court and in 1897 its decision reversed that of the lower court. The new plant, having been a bone of contention for three years, started in 1898. In the following winter complaints were made that the motors did not work satisfactorily and that another expen-

with that of 1902, fell off over 40 per cent. These figures clearly show that industrial friction was the direct cause of the increase of socialistic sentiment in our territory, and that the restoration of peace and prosperity has arrested its progress.

diture of $2,500 or $3,000 was necessary before all would be right. By this time the patience of the people was exhausted, but politicians are notorious for their persistency, and, in 1903, the above sum was appropriated. Thus in eleven years, during two of which no electric light was enjoyed by the people of the borough because of litigation, over $26,000 was spent on an electric plant. It has never run smoothly although in the six first years of its life the sum of $580.64 was spent on oil. The town has never known how much has been spent for tools and supplies, for some of the secretaries had the habit of destroying the account books, if ever they kept any. Hammers and wrenches, chisels and cans, disappeared as if they had wings. A $38 water meter was stolen before it was paid for, and the material in the foundation of the old plant was carried away piece-meal without a word of protest from the council, for some of its leading members were the chief transgressors. The council employed the men necessary to run the plant, who invariably reflected the change incident to public elections. Men were chosen, not because they were capable and sober, but because they were cronies of the ruling faction. As long as their friends dominated the council they had no boss. There was no one to enforce discipline when they grossly neglected their duty. The plant was not cared for as that of a private corporation would have been, and although the council monopolized the selling of globes and the secretary alone knew how many of them were in the borough, no accurate account of this item of expenditure was ever given the public. All the firms dealing in electric goods east of Chicago were drawn upon and so bad was the credit of the borough that no goods were forwarded save on the C. O. D. basis. For weeks at a time the plant remained idle for the want of supplies and with no money on hand to procure them.

From 1895 to 1900 the borough has been involved in 17 suits arising from the attempt to build an electric plant. There is no way of knowing how much money has been spent in these suits, but the following bills recorded will give some idea: Sept., 1895, $100; Sept., 1896, $343; Sept., 1897, $355;

March, 1899, $491.50; a total of $1,289.50. Municipal ownership of public utilities to the above borough has simply resulted in wastefulness, inefficiency, nepotism, bribery, litigation and scandal, and many of the people now say "sell the plant and end the dirty work."

The borough of Shenandoah in 1895 decided to construct a water plant to supply water to the borough. For two years the private corporation already in town impeded the progress of the design and offered to sell its plant to the borough for $100,000. This offer was rejected and in 1897 the citizens got the right to construct a plant of their own. They invested $150,000 in one. In connection with the plant was a receiving reservoir. This, soon after its construction, leaked. The service of an expert was secured who said it would cost $42,000 to make it water-proof. The money was spent and yet the reservoir leaked. Another expert was engaged and his report was that $38,000 was necessary to put it in order. That money was also spent and in January, 1902, another defect was reported which cost a few thousands more to make good. This money has again been spent and, this fall, the officers are afraid to put more than eight feet of water in the reservoir, and have cut down the supply to 12 hours out of every 24. It is claimed that a total investment of nearly $350,000 has been made and yet in the months of summer, when water is most needed, the mining population of Shenandoah is put on two and three hours' allowance of water every 24 hours, and 90 per cent. of the homes is supplied by this plant. The receipts from the plant in 1901 was $14,828.80 and the cost of running it $13,547.90; in 1902, the receipts were $12,889.38 and the running expenses $17,-895.21. The borough pays 5 per cent. on $209,500 bonds, most of which were issued to secure money for the work. It has a competitor in the private plant, which accounts for the rates being 25 cents a month for private houses and 50 cents for saloons. Last April, the rates were raised to 50 cents a family, but the increased cost of fuel for the pumping engines will consume most of the increased income. Thus, although the citizens have had a plant of their own, they pay interest and principal by taxation and in the summer months suffer an inconvenience that is little less than criminal in a mining town.

In the construction of the plant there seems to have been a lack of engineering skill or judgment, else the money spent would have given better results to the people of this flourishing town. The supply of water in the water-shed is abundant; the fault is in the foresight of the engineers to plan to meet the requirements of the borough every season of the year.

The saloons and business men of the town get a supply of water from the borough plant far in excess of the rates they pay if compared with the amount paid by private homes. When we asked one of the officers why they could not adjust rates more equitably, he said: "Political influence will not allow it." The politicians are afraid to offend the saloonists and the business men by a more equitable adjustment of rates. Thus the borough is mulcted annually to pay principal and interest because of the selfishness of business men and the timidity of politicians. Here again, all laborers connected with the plant and residing in the borough are changed when the political complexion of the council changes. The engineers and firemen who operate the pumps, which are several miles from the borough, are not changed.

Other enterprises undertaken by the municipalities in the interest of the people may be summed up in the same way: inefficient service, waste of public funds, political intrigue, nepotism, litigation and vexation of spirit.

How can it be otherwise? The average citizen enters office, not to serve the public, but to seek his own interest. The moment a borough undertakes a public enterprise the councilmen are besieged by a score of agents looking for a job and the most unscrupulous offer the largest bribes. The men in our councils are for the greater part employees in and around the mines and know not the first thing about business or works of public utility. Providing they are honest, how easy it is for men of this stamp to fall victims to shrewd agents, while the unscrupulous official will follow the scamp who places the largest sum in his itching palm. What does a miner know about an electric plant, or a macadamized road, or a water plant worth $350,000? Reliable and well-established firms have their price and no boodle; unscrupulous firms that send their agents to saloons to fix councilmen have their prices and boodle, and nine cases out of ten, with the composition of our councils being what it is, the unreliable gets the job.

Our people often complain of the high rates charged by private corporations doing business in boroughs. How can it be otherwise? A percentage of their earnings must ever be laid aside to satisfy parasite councilmen who abuse the trust reposed in them. Street railways and steam railroads must annually give passes to councilmen, saloons must give them free cigars and beer, and private citizens who have any favor to ask must indulge their sensuous appetites. A corporation that comes into town to do business must first of all be "plucked" by a crowd of thirsty knaves which ever look for prey. A company in one of our boroughs was periodically blackmailed by sinister members of the council that it may escape harass from threatening municipal restrictions. To meet these pilferings of the profits by the servants of the people, the private corporation must charge a proportionately higher rate for the service it renders. On the other hand, attempts made by the people themselves to escape the exorbitant charges of private corporaations, result in inadequate service and higher taxes, for again the politicians are there feeding on the public body. In the work of both private corporations and municipal enterprise the secret of exorbitant charges lies in the corruption of public servants, and the curse will never depart from our municipalities until the people themselves intelligently and conscientiously choose men of unquestionable integrity and sterling honesty to serve them.

CHAPTER XIII.

"THE CONCLUSION OF THE WHOLE MATTER."

1. The Ills We Bear: (*a*) In the Homes and in the Schools; (*b*) In Society and in the Works; (*c*) In the Boroughs and in the Counties. 2. The Way to Health: (*a*) Through Harrisburg; (*b*) Through Personal Effort; (*c*) Through the Church.

The Ills We Bear.

In the Homes and in the Schools.

We saw in the first chapter that the increase of population in our area for the decade 1890–1900, was over 6 per cent. higher than that of the State in general. This was largely due to the influx of Sclavs and Italians as well as the high birth rate among the prolific "foreigners." Few have been the immigrants from Great Britain or Germany into these coal fields in recent years. The Commissioner of Immigration said recently that 70 per cent. of the immigrants of late years intended to settle in New York, New Jersey, Pennsylvania and Massachusetts, and that the percentage of British and Germans among these was insignificant. This shows that the available supply of cheap labor is unlimited, and when George F. Baer, President of the Reading railroad, said before the Coal Strike Commission, that the excessive supply of immigrant labor about the mines was the result of the abnormally high wages already accorded and that the excess would be increased if wages were again advanced, he touched upon a law in economics which labor unions will be helpless to check. And as long as this stream of cheap labor will flow into our country, these anthracite coal fields will be invaded, for labor in and around the mines, with the present rate of wages, will be a lure to the unskilled immigrants from Southern Europe.

The result of this influx is three-fold. (*a*) The better class of labor is pushed up and out by the cheaper grade, (*b*) our society feels the effect of intensive evolution, and (*c*) undesirable elements enter into our social group. The English-speaking element of our population feels the pressure from beneath and those who do not leave the coal fields for the cities soon find it necessary to abstain from marriage in order to maintain the change due to a rising standard of living. Hence our towns and villages are being depleted of the better class of citizens, and many of those who still remain prefer ease and amusement to struggle and responsibility. These are ambitious to rise in their social status, but they are not willing to pay the price their fathers did in hard work and self-denial, and hence they sacrifice quantity for quality. All our population also feels the effect of an intensive evolution. Lilienfeld says that a head with 1,000,000 nervous cells will function less intensively than one with 2,000,000. The same is true of an area that is sparsely populated and one that is crowded to the limit of its means of subsistence. A population of three quarters of a million, subsisting on the production of these collieries, means a stronger and more intense life than was known in these coal fields a decade or two ago. The influence is felt in every sphere of life. Business is more competitive, individuals conflict more frequently in the mines and in commerce, the public interests demand closer attention and the members of society are more precisely differentiated into groups arranged in a hierarchical order. Rights and duties are more precisely defined, police regulations and organization must be of a higher type in order to force the recalcitrant elements to do their duty and obey the laws. Private property is not so secure and there are demands for greater vigilance on the part of constituted authority to preserve it, while the rights of citizens are not so religiously respected. Those who have watched the growth of our population have observed these changes, which work for degeneracy in our communities and are coincident with the increase in population. But not only do we suffer from intensive evolution, the character of the elements added to our population has also much to do with our present conditions. While we readily confess that the Sclav contingent of strong men, in the prime of life, has greatly facilitated the development of our industry, we also feel that these communities need something more than material development. We need men and women who harmonize with national ideals and character, which is of greater importance than pecuniary gain. The thousands of immigrants added to our population have lowered our standard of living, have bred discontent, and have brought elements that are utterly un-American in ideas and aspirations into our communities. These, by their adherence to their language and customs, remain unassimilated after years of residence in the United States. This works disintegration in our industrial and social life and, unless counteracting forces are set in motion, will result in a lower type of manhood and womanhood in these communities.

In the sphere of the home, among the majority of mine employees, there is much to be desired. The great need is better mothers. Mothers who know how to care for their children, prepare food for them, and understand the importance of cleanliness, fresh air, pure water and sunshine. Wives who know how to cook and the relative value of foods, how to mend and darn and the importance of fresh air and cleanliness in the home, how to care for the sick and the use of disinfectants, how to direct the affairs of the home and watch over the work necessary for the comfort of their families. These are the qualities needed in the wives and mothers of our mining population, and the reason why so much misery exists in our homes is because the mothers and wives are ignorant of the domestic arts. It cannot be otherwise as long as girls go to factories and stores, and are, as their mothers, ignorant of the work of ordering a home and the arts of home consumption. The remedy will not come until these arts are taught in the public schools, and our girls, instead of being loaded down with studies they forget soon after graduation, are trained in plain cooking, dress-making, washing and ironing, etc., which are daily in demand.

This is the more important for the wages upon which the family must live are often inadequate to meet ordinary demands. It was shown by the Bureau of Statistics of Massachusetts that it takes a family of five persons $754 a year to live on. The average number in the family of mine employees is between five and six, and the wages of contract miners, who form only twenty-five per cent. of all persons employed in and around the mines, is about $600 a year; while adults in other classes of mine workers, who form over sixty per cent. of the labor force, do not receive an average annual wage of $450. On this income, it requires the greatest possible skill to provide for the bodily necessaries of the household, and leave nothing for the "spiritual affairs of life—those affairs that are above and beyond the mere contest for subsistence." This small income drives many of our people to live in cheap and rickety houses, where the sense of shame and decency is blunted in early youth, and where men cannot find such home comforts as will counteract the attractions of the saloon. The hundreds of company houses, renting for from $1.75 to $3 per month, are not fit habitations for men; these should be torn down, for self-respect and decorum cannot be cultivated in families that live in them.

Another evil incident to a small income is the contracting of hopeless debt. This has a positive deteriorating influence upon the family. It lessens the moral tone of the members, makes the parents increasingly careless of obligations incurred, and affects the community in creating distrust and suspicion of the integrity and honesty of its members. Whenever it is necessary for a family to incur debt, in order to secure the necessaries of life for its members, the effect is detrimental to the welfare of the home and deranges social relations. Ideas govern the world, and when children are raised in homes where the concepts of truth and honesty are perturbed because of economic conditions, they become disturbing elements in society. Much of the social unrest which prevails among the working classes arises from the idea that they fall into debt because the distribution of the wealth produced is not equitable.

Economy in a Mining Town. (Twenty-five families are supplied by this hydrant.)

Whether this be true or false, the effect upon society, should the idea become general among the wage earners, must be to destroy its peace and introduce a disturbing element into all the spheres of human life.

In all the homes of mine employees, of all nationalities, is an appalling infant mortality. Hundreds of children in these towns and villages die every year of preventable causes. While not five per cent. of the children born to the upper classes dies before they reach five years of age, the death-rate here is about 35 per cent. This is a frightful infant mortality and should be the concern of the philanthropist and economist. The causes of this waste will be found in the facts that the social and economic conditions of these peoples are low; that their knowledge of the care and treatment of infants is small; and that the vice, immorality and debauchery of unnatural parents enfeeble the constitutions of their children. Life in all its stages is a struggle and the law which a large percentage of our people seems to be wholly oblivious of is, that the physical constitutions of men and women depend ultimately upon morality. The diseases which men contract because of immoral habits and practices account, by the laws of heredity, for the weak constitutions transmitted to descendants, who thus become a prey to adverse influences and succumb. Whatever knowledge is necessary to check this frightful waste ought to be furnished from considerations of economy and morality, for the growth of a healthy, intelligent and vigorous laboring population is the concern of society and the only basis upon which civilization will thrive. A decline in physical power connotes a decline in moral force, and a decline in morals means the decline of the nation. Ultimately the student of society comes face to face with the fact that the higher welfare of the nation depends upon the welfare of the working classes.

And can this welfare of the working classes be promoted in any more effectual way than by an intelligent system of education, whereby the best use is made of the predilections of youths? A recent examination was conducted by one of our Factory Inspectors of the 2,000 children employed in the factories of Lackawanna county, whether they could read and write the English language intelligently, as if that were a sufficient equipment for the conflict of life. Statistics show that 50 per cent. of the children of the working classes is out of school before they are twelve years of age. The cause for this is not so much economic as it is a conscious or unconscious protest of the common sense of the masses against the system of education which now prevails for boys and girls in their teens. Gibbon says that the modest obedience of the legions of the Roman army, during the two first centuries of the Imperial history, was due to the men who were instructed "in the advantages of laws and letters, and who had risen by equal steps through the regular succession of civil and military honors." Rome adapted its system of education to the exigencies of the times. Can we do anything better? We live in the industrial age, and over 95 per cent. of all our boys have to eke out a subsistence by manual labor. Does not common sense dictate that these youths should be trained to take their place in the industrial army in a more efficient manner than is now the custom? The boys, who have a smattering of a dozen different sciences and have their certificates of graduation hung up in the homes, are very poor material for manual labor in these mines. Send them into the city and what good are they there? They soon sink into obscurity and darkness and, unobserved by any who takes an interest in them, they are very likely to neglect themselves and in time abandon themselves to every sort of profligacy and vice. We know that the dream of certain socialists of a society where all inequalities are leveled, is contrary to both nature and reason. Differences of native capacity will always exist and there will always be great men whose social worth is inestimable. But while all intelligent men accept a social state in which differences among men is normal, they will, with equal discernment, perceive that the peace and welfare of society rest upon giving every youth equality of opportunity to develop his native powers under competent masters in the industrial world. The councils of Wilkesbarre have become conscious of the deterioration of the youth of that city and have instituted the "curfew bell." We need something more than curfew bells to check the tendency of precocious youths to vice and crime. We need industrial and technical schools wherein the nimble hands and fingers may be properly trained and the surplus energy of youth turned to profitable channels.

In Society and in the Works.

There are many pathological conditions also found in our social and economic relations. We are now on a wave of prosperity and all mine employees are elated with the increase of wages granted them. Business is booming, the collieries are working regularly,* the volume of consumable goods devoured by our people was never greater, but no one seems to ask what is the character and utility of the articles consumed. What will increased wages benefit men if they consume them in goods of neutral or negative utility? Saloons, tobacco stores, cock-fighting matches, vicious shows, and gambling devices flourish, while good books, refined amusements, culture and religion are very meagerly patronized. The way people spend their money is a true indication of the strength or weakness of society. Intelligent men imagine that everything is prospering because the volume of business is great, and perceive not that the volume of business may indicate the presence of anti-social and anti-patriotic forces which work for the destruction of society. Business that destroys the physical and moral fiber of men is a sure sign of deterioration.

During the recent strike in these coal fields, Lieutenant J. P. Ryan, of the United States' Navy, was sent here to recruit sailors for the navy, and then ordered to leave the regions for the following reason: "One curious outcome of the recruiting was that very few of the strikers who applied for enlistment could pass the physical requirements. Lieutenant Ryan found that nearly all of those who were willing to go into the navy are under size, weak-chested and round-shouldered, and phys-

* After a year of unparalled prosperity since the strike of 1902, intermittent labor is again the lot of anthracite employees. The collieries do not now average more than two thirds time.

ically undesirable in nearly every way. This is attributed to going to work at an early age, lack of nourishment and hard labor in unsanitary surroundings." Persons familiar with our regions know also that the intemperate and vicious habits which prevail among the lower stratum of our society account for much of this result.

Take the evil of alcoholism in our regions and it destroys both the body and the souls of men. Dr. G. Sims Woodhead lately said in discussing the pathology of alcoholism, that alcohol induces cirrhotic changes in kidney and liver, that pneumonia in alcoholic patients assumes the most virulent form and often terminates fatally, that it interferes with nutrition, and predisposes the indulger to infection. Colonel H. M. Boies, writing from personal observation and from the record of crime in the courts of justice in these coal fields, said that intemperance is the cause of 75 per cent.* of the crimes committed. And from the coal operators come the complaints that the number of days lost by habitual drunkards in the collieries amounts to about 8 per cent. of the time worked by the breakers. This evil causes most serious disturbances in the social, judicial and economic spheres of our society and is one of the most serious dangers which confront our people. It unsettles our economic life, destroys the peace of families, deprives children of their natural rights, brutalizes and debauches the moral life, and leaves us a heritage of pauperism and crime, diseases and insanity, degeneracy and suicide. Against this great evil both the economic and moral interests of our communities should protest and wage war. Col. Carroll D. Wright has said: "Drunkenness and intemperance are not the necessary accompanying evils of the factory system, and never have been; but wherever corporations furnish unhealthy home surroundings, there the evils of intemperance will be more or less felt in all the directions in which the results of ruin find their wonderful ramifications." The evil has ever been great among mine employees, and possibly the nature of the industry and the conditions of domestic life have much to do with the curse.

* This is a higher percentage than that we found in typical mining towns. See note on page 286.

Closely associated with alcoholism is crime. The estimated number of crimes committed in the country in 1900 was one for every 22.13 of the population. This is about the proportion we found in our area. These anti-social elements are a burden to society that handicaps its progress and disturbs its peace. The number of criminals increases and possibly will continue to do so until a scientific and curative form of treatment is applied. From the days of Mabillon, the Abbe of St. Germain, to those of Z. R. Brockway, the leaders in penology, have advocated "reformation" as the motto in the treatment of criminals. Yet, at the beginning of the twentieth century, we hand over our transgressors, regardless of age and antecedents, to the care of politicians in charge of county jails, notwithstanding the protests of specialists such as Eugene Smith, Drähms, Boies, etc. Is it any wonder that criminals multiply and crime increases? The last mentioned authority says that 75 per cent. of all crimes committed and at least 50 per cent. of all sufferings endured on account of poverty in this country and among civilized nations are due directly or indirectly to alcoholism. Other causes are inefficiency or absence of parental discipline in early life, lack of self-control, idleness and disinclination for steady and methodical work, and moral depravity. These sources, whence come our criminals, cannot be cleansed or removed until the positive deductions of the science of penology are made the basis of intelligent action. When that is done the causes will be eliminated and the criminal type, which closely resembles the barbarous and savage type, will be largely removed.

Our pauperism is also closely related to alcoholism. If we consider the economic and social conditions of these coal fields there is no reason why the large sums spent in poor relief should be as great. In the treatment of the pauper we are far from following scientific methods and principles. Our volunteer charitable organizations do not always bear in mind that their aim should be to redeem and cure and not help inconsiderately and continuously. The men in charge of poor relief have no idea that they are dealing with a disease which needs to be stamped out. Their reports read as those of organizations which exist for purposes of positive utility and each year shows a larger volume of business. When public-spirited citizens have taken this question in hand and acted in the interest of the public, the list of paupers has been cut down one half without working any injury to the worthy dependents.

We cannot regard otherwise than as abnormal the craving for excitement and the tendency towards extravagance observed among our working classes. No one will object to a healthful rise in the standard of living, but when taste and culture do not keep pace with the increased wants felt by the wives and daughters of wage earners, their taste becomes grotesque, their costume preposterous and their homes repulsive with gaudy tapestry and cheap upholstery that are fatal to simplicity and virtue. Among the wage earners there is an unnatural rivalry for display. Their idea of social preëminence is to make an ostentatious show in clothing and furnishing, and waving over empty heads and vain hearts are seen costly plumes, ribbons, rosettes and silks which represent a depraved taste and false ideals of life. The Sclavs, many of whom are graceful and natural in their bright colored scarf and simple gown, are now mimicing their Anglo-Saxon sisters and, with flamboyant hats, silk waists and fashionable gowns, they present a more ludicrous picture than ever entered the mind of Thackeray or was caricatured by Punch. If coincident with the desire for a greater share of the wealth produced, there was a desire for a higher intellectual life and a greater realization of the moral and spiritual affairs of life, one would look upon the advance of the wage-earning classes with greater hope.

Closely related to this craving for material enjoyment lies the growing prejudice against the capitalistic class and jealousy of the wealth it possesses. The grossest forms in which these ideas are found are certain types of socialism; while many who are not fascinated by socialistic dreams of "paradise regained," are still suspicious of the share of the productive wealth which goes to capital. Adam Smith said, "It is the interest of every man to live as much at his ease as he can," but the working classes, if they fall into the delusion that the capitalistic system alone stands in the way of the coming of the kingdom of ease and plenty, will be rudely awakened from their delusion. The work of getting food has always been difficult and never in the history of the world has there been such a supply of useful commodities for the use of man as under the capitalistic system, while those countries where vast aggregations of capital rest in the hands of few men lead the world in commercial and industrial efficiency. Natural endowments are unequal and the returns each man receives for his labor under our system represents, in a rough way, the social worth of the parties. Native capacity inexorably fixes the limit of every man's achievements and never will society esteem the services of the ashman equal to those of a railroad president. The foolish and extravagant invention of parasitic millionaires who furnish dinners on horseback at the rate of $200 a plate as well as the pinching poverty of laborers whose annual income, in a country where revenue for monopolies prevails, does not double that amount, may prepare the way for socialism ; but the old adage is still true, "two blacks will not make a white," and the tendency of the working classes to wage war upon the capitalistic class bodes no good for the peace of society. The antagonism between capital and labor is daily increasing and the mutual confidence and sentiment of solidarity, which are the *sine qua non* of prosperity, are weak. The gulf between capital and labor becomes daily wider, but it cannot be spanned by either communism or socialism. The demand of the workingman for a wage sufficient to enable him to meet his social and spiritual wants beyond the necessity of keeping body and soul together is just, but we believe that other ethical and economical adjustments can be found to meet this demand without launching on the uncertain sea of socialistic experiments.

More and more does modern society feel that ethics and economics must go hand in hand. Capital and labor are mutually dependent the one on the other, and when each consults the other and reciprocal interests prevail, when a moral standard actuates both parties and the egoistic principle is subordinated to the altruistic, then will the best prosperity be attained and sustained in our communities.

The last and most important pathological feature we need mention is the moral decadence that is so apparent on all sides. Wayne MacVeagh said : "For it may well happen that the safety of our institutions requires that the masses of our people shall continue to cherish the ethical ideals of Christianity, and that whoever lessens respect for them inevitably weakens the reverence of the majority of voters for the principles upon which our government is founded." The trouble is that there is a weakening of reverence and a growing disregard for sacred things on all sides, that threaten the very foundations upon which society has thus far rested. Charles F. Dole touches the disease when he says, that "There is seething unrest ; there is doubt of the sanctions of religion ; there is a sense of coming change ; there is suspicion that premises and foundations once unquestioned are now perhaps undermined ; there is challenge of existing institutions — social, economical, ecclesiastical. Are the present institutions such as the world will continue to find use for ? There is dread mingled with hope. What possible revolutions may not impend, setting the old order aside." And while this unrest and distrust are in the air the masses will not continue to cherish the ethical ideals of Christianity. Cardinal Gibbons speaking last February said : "There is a barbarism more dense than the barbarism of the savage. . . . I speak of the barbarism which eliminates God and an overruling Providence from the moral government of the world, which takes no account of a life to come and of the responsibility attached to it."

This decay of religion that is observed and commented upon by our religious and secular leaders is ominous. Are the lessons taught by pagan Rome, by the reigns of Louis XIV. and Louis XV., by the reigns of Charles I. and Charles II., and by subsequent periods, when material prosperity declined as the practical resultant of immorality and profligacy, in vain? Or are the melancholic predictions of Frietschke and his class to be realized, that the vast majority of mankind are only capable of rising so high and, then falling, they bury the achievements of ages in one great catastrophe, from which man must again painfully and slowly arise only, in time, to repeat the calamity. If society is to be saved the regenerating power can only come from the moral-spiritual nature of man, and every force, either in society or industry, which grinds the altars of the nation will ultimately grind to powder the foundations upon which society rests.

In the Boroughs and in the Counties.

Every part of our State suffers from the corruption which characterizes the actions of politicians. One of the most humiliating spectacles was presented to the people of the State during the Quay-Elkin fight of 1902. Men were shipped like cattle from the anthracite coal fields to Harrisburg to shout and fight for their patron. Never were the contests between rival candidates for the imperial crown in the decadent days of ancient Rome more venal, corrupt and barbarous. The state administration used all the patronage at its disposal to aid its favorite, and those who would not obey were summarily dismissed. It is hardly credible that such a spectacle could take place in a self-respecting community, but in every part of industrial Pennsylvania the same corrupting and deadly methods are pursued by professional politicians. There is no sphere of public life exempt from their mischievous touch and no weapon is too vile for use to accomplish their end. The idea is daily growing among the working classes that the law and law-makers is only a question of money, and that the right to make laws is to be purchased as one does any other commodity. Daily it is becoming more apparent that all the machinery of law is at the service of those who possess wealth enough to control it. The sentiment that the question of law is a question of morals is antiquated. Capitalists have openly and cynically taught the masses that it is a question of hard cash and that the legislature can be run on schedule time according to the pleasure of the men who have the "barrels." It is a mystery how intelligent men can condone such a state of affairs and how men, prominent in church affairs, can warp their consciences so as to connive at such political corruption.

The deadly blight of politics falls on all our public institutions. Our schools are wholly in their hands, and teachers are appointed because of influence and not efficiency, janitors must have a "pull" or else be dismissed, and school supplies must be bought at the hands of men who can hardly read or write the English language. Almost all the public officers of cities, boroughs and townships are politicians and belong to the machine. They secure their elections by ways in vogue among men who conduct campaigns on a larger scale, and, when in office, they are anxious to get all "that's in it." Egoism is the motive power of their lives and altruistic motives for the public good are far removed from them. Charities and sanitation are in their control; the poor tax they distribute recklessly and just as recklessly leave nuisances unabated. Col. Boise, in speaking of the provision for the indigent of Lackawanna county, says: "Five institutions with five farms, five superintendents all with their hired staffs, are all employed to accomplish what could be done in a much more satisfactory manner in one institution, on a single farm, with one superintendent." Politicians never reduce expenses and help; they increase both, for that enhances their chance of continued official life. That is the reason why the cost of governing our communities has increased much faster than our population, while the burden of taxation which falls upon the thrifty and industrious wage earner has steadily grown heavier. Even the judiciary is not free from the rampage, scramble and smudge which characterize our political contests.

The influence of such a disintegrating force is appalling upon the political ethics of the people. One of the most direct results is the organization of political clubs along racial and religious lines which excites prejudice and invites corruption. Adam Smith said: "The good temper and moderation of contending factions, seem to be the most essential circumstance in the public morals of a free people," but the political corruption which characterizes our communities inflames passion, in-

Types in a Mining Town.

tensifies hatred and, if it long continues, will precipitate bloody conflicts between rival factions in quest of the spoils of office. Unscrupulous schemers, in all our towns, organize clubs of their own people, and, posing before the party leaders as able to control a large number of votes, put them up for auction or barter them for some personal favor. All this is un-American and contrary to the spirit of our institutions.

Another result equally apparent is the decay of public morals. Men that are honest and truthful in social and business affairs are dishonest and unreliable in political affairs. In private affairs they give good service for all remuneration and would never think of taking a dollar that was not their own, but in the service of the public they skimp work, make excessive charges, and hesitate not to take from the public treasury money for which they gave no service. The effect upon the character of the politicians themselves is ruinous. They live in an atmosphere of cunning, intrigue, venality and prevarication, and its demoralizing influence, in time, tells upon all their activities. They introduce the cunning and craftiness which are practiced in politics into business, their intimate friends are those whom they can use, the sacred ties of family life are sacrificed on the altar of political ambition, personal character is abandoned, and their church connections are only maintained to throw a mask of decency over their perfidy before the eyes of the public. It is surprising that keen business men, who accumulate wealth and hope to transmit it intact to their descendants, do not see that this destruction and denial of ethical ideals in a democratic form of government tends to destroy all that is sacred to a free people. The common people have always followed the lead of the rich, and to-day the masses in our communities are honeycombed with false and fatal ethical principles in politics, which they have taken from the mighty.

This kind of politics has brought to us an heritage of misappropriation, incapacity and extravagance in municipal and county affairs that is ominous to the public weal. Every piece of work costs from 25 to 50 per cent. more than a private party could get it for. Public-spirited citizens, who try to check the extravagance, are confronted by the machine which shields its minions and defeats the ends of justice. Men who are known to have misappropriated public funds, are exonerated if only they replace the money and be more careful — to cover their tracks — in future. Men are paid high wages in county offices whose only qualification is that they can command votes and, when necessary, lay their hands to work that no honest citizen would touch. And intelligent people, who know all this, are now talking of enlarging the sphere of operation of these political freebooters by municipal ownership of public utilities; and even our legislators, ever sensitive to public clamor, are introducing laws which pave the way to socialistic experiments, which, they know full well, must result in deeper corruption and greater extravagance than exist at present. Are these men blind to the fact that State interference with the enterprises of its citizens will inevitably destroy the sentiments of initiative and responsibility which have been the crowning glory of our Commonwealth? The few experiments in municipal ownership of public utilities in our territory have increased the burden of taxation, intensified factional feeling, and led to expensive litigation and scandalous extravagance. But these evils are not the worse, for as M. Le Bon has shown, the psychological effect of this drift of public sentiment is far more injurious to society than its increased cost. It degenerates the individual will and energy, and ends "in a kind of bureaucratic servitude or parliamentary Cæsarism which will at once enervate and demoralize an impoverished country."

The Way to Health.

Through Harrisburg.

Kent says that municipal law is a rule of civil conduct prescribed by the State which must be reasonable and conformable to the will of the people as expressed in the Constitution, which is the supreme power in the land. The civil code of all countries embraces the laws which reflect the social limitations and conditions which society has gradually built up in order to cause its members to observe propriety in all their acts. The laws do not always respond to the demands of right and justice, but, taking the evolution of social justice as a whole, we find that they represent honor, justice and equity which gradually rise above the instincts of brutes and the caprices of savages. The horde fixed the relation of the member to the group, and, when by custom this was established, the mores of the people were formed, which, in turn, crystallized into laws. Hence statutes are only the reflection of the mores of society and a secondary factor in the judicial life of a community, but the sanction they receive from the regularly constituted authority of the State gives them considerable influence in molding the life of the people. Thus it is that laws, as they are beneficent or mischievous, become instruments of progress or destruction in the life of society.

Laws, when they are beneficent, promote life and happiness. They are means designed to eliminate impropriety and mischief on the part of the individual who works injury to his neighbor. So that all the machinery of government, designed to benefit society, reaches its aim only as the individual is corrected, and where all the individuals of a group are in equilibrium with the social code, they are a law unto themselves. But as Frederick Harrison says, "The devil of separate interests drives man apart from man," and the coercive power of the State is evoked by the aggrieved to keep within bounds the devil of separate interest which commutes the welfare of the masses to its own purpose.

The trend of modern legislation is to curb the strong in the interests of the masses. The aim is to equalize men in all the social spheres by coercion. This tendency is diametrically opposed to the results achieved by the free play of social forces under the lead of glorious liberty. Whatever may be the outcome of this spirit of the age, it must always be acknowledged that the policy of individual initiative and responsibility formed men capable of achieving great actions, of laying the foundations of great empires, and in no quarter of the world are there found such a galaxy of noble men as are found in countries where individual will and energy were given free play.

All students of social science acknowledge that individuality and solidarity are the two poles around which social life revolves, but they differ as to the part government should play in the adjusting of these forces. Every Christian believes that we are members one of another, that the interests of all men and all nations are identical, and that true progress and true prosperity for all are to be found in universal mutual service, which can never be effected by law. The harmony and unity of nature are not realized by the laws of nature, but by conflict among the various elements of the organisms, and society will best flourish when free play is given social forces and when governmental activity is confined to laws designed to secure each member equality of opportunity and equality of rights in the conflict of life. The words of John Fiske, which express the conclusion he drew from the study of paternalism in New England, deserve consideration: "For while it is true —though few people know it—that by no imaginable artifice can you make a society that is better than the human units you put into it, it is also true that nothing is easier than to make a society that is worse than its units."

It is well for our legislators also to know that the multiplicity of laws passed will not guarantee the progress of society. The Justinian Code did not save the Byzantine Empire from economic and political ruin. It were better for us if fewer laws were passed and greater thought given to secure statutes which mean social advancement. Politicians laugh at students of social science who set forth the need of society. Rude barbarism in times passed hindered the advancement of science by scorning and persecuting the students of their days, and blatant demagogues swayed by the same motives, still scotch the progress of society. Before the noisy existence of the industrial age began, the sedentary, quiet, thinking people had done their work and prepared the way. And the wage-earning classes must learn, before their future well-being is placed on a sure foundation, that the healers of ephemeral popularity, whose vigor lies in their voices, can never, by inadequate and dangerous legislative remedies, cure their diseases. Time will prove that their loud phrases and chimeras are only hobby-horses on which popular heroes ride for a day.

The greatest statesmen the world has seen have only been able to direct and regulate the vital energies which repose in the bosom of society. Solon could only establish the laws the people could bear and not the best system of laws he could devise. It is only up-start statesmen who are confident that they can change society to order, as the potter the clay. They forget that the material upon which they work is an association of psycho-physical energies and not an inorganic mass governed by mechanical principles and independent of physical environment. If our legislators were trained in history and social science they would know that continuity is one of the laws of evolution and the necessary condition of existence, and that the scalpel ought not to be used in human affairs save when all other remedies have failed. Cicero says that the divine maxim of Plato was, never to use violence to his country any more than to his parents, but our ready-made legislators, who are picked up from the street or taken from the saloon and sent to grind out laws, attempt suddenly to destroy the historic type of our community and precipitate disastrous consequences. The clear-sighted David Hume said: "In all cases it must be advantageous to know what is most perfect in the kind, that we may be able to bring any real constitution or form of government as near it as possible, by such gentle alterations and innovations as may not give too great disturbance to society." When the working classes will understand these principles, they will not be hypnotized by blind leaders whose mischievous interference abases society and leads the unconscious throng to hopeless difficulties.

There are lines along which intelligent legislative action can exalt our society. Population is increased by the high birth rate among our working classes. The investigations of Engelmann above referred to and those of President Eliot prove that the middle and educated classes fail to reproduce themselves.

That is not the case here. Large families are the rule among the majority of households in these coal fields, but the rate of infant mortality is appalling. This is largely due to preventable causes and offers a field of action to wise legislators to render patriotic service. Let the sanitary laws be executed and rigidly enforced, and let parents who allow their children to sicken and die without making an effort to secure the service of physicians, feel the effect of stringent laws covering the neglect. It is also little less than criminal to force many of our people to live in habitations that do not keep out the cold of winter which frail infants cannot stand. If the companies will not tear down the rickety old shacks from which they have derived an income for the last fifty years or more, then compulsory legislation should be brought to bear upon them to provide better dwellings for their employees.

The laws regulating child labor should be uniform and better executed. The State of Pennsylvania is behind many of her sister states in this, but the recent agitation, started by the revelations made before the Coal Strike Commission, bids fair to effect an improvement as far as the anthracite regions are concerned. We can never hope to assimilate the mass of raw material in these coal fields unless we insist on their enlightenment. Every patriot should insist upon giving every child the rudiments of knowledge which will enable him to possess that degree of intelligence which may reasonably be expected in a citizen of a democracy. Provision should also be made by law for the establishment of kindergarten schools in every school district in our territory. Dr. Eugene Smith says: "Science proclaims . . . that every dollar that the State expends in providing that wise and uniform nurture of its children which they need before reaching school age, is worth ten spent after that in correction and education, and a thousand expended for protection from criminality and the reformation of criminals." Dr. Harris also says of these schools: "It is a part of the system, as an adjunct to the public schools, to educate every woman in the valuable matters relating to the early training of children."

Some means should also be devised to take our educational

system out of the hands of politicians. As affairs are now conducted the locally elected boards of directors are supreme and the State exercises no supervisory power whatsoever. The results are inefficiency in the schools; defective plans of instruction; lack of uniformity in studies and methods so that pupils transferred from one borough to another cannot continue their studies and are often discouraged and leave school; corruption in business transactions and the selection of directors, who are dominated by egoistic motives, and wholly incapable of rendering the best service to the wards of the State. Patriotic sentiment should make such a condition as this impossible, and the rights of the rising generation should be better and more efficiently guarded. The education of our youth ought to be entrusted to the hands of trained men, who should give the State a uniform system best adapted to the requirements of our youth, and wholly removed from party politics.

It would seem also that the training of boys and girls for the business of life requires a modification of our present educational system. The rudiments of a common education ought to be acquired by every normal child by the time he is twelve years of age; after that, those who, by social status and native capacity have no prospect before them other than manual labor, ought to have technical instruction whereby they may be equipped for useful service in the industrial army. It seems an anomaly that delinquent children should be better treated in this regard than those of regular and orderly habits. Mr. O. Eltzbacher discussed this question in the February (1903) number of the *Nineteenth Century* and said: "Let us hope that the spirit of combination which seems to be growing, though somewhat slowly, within the community, will in due course dot the whole country with technical schools founded and supervised by the various industries themselves and planted under the eye of those industries in their business center. . . . Let us hope, besides, that the direct active interest in education, which practical men are beginning to take, will cause in course of time the mapping out of specialized school programmes by competent experts for all schools from elementary schools to universities throughout the country; for, after all, practical men, not tradition-bound schoolmasters and well-meaning clergymen, can determine the practical requirements of education."

The female youths should also be provided for by instituting schools where each could be trained in the duties of motherhood and in the domestic arts. The hope of the world for a healthy, vigorous, energetic, and conscientious class of working people lies in a better type of motherhood, and the best investment the State can make is to build up a higher grade of mothers — women who possess knowledge and intelligence to know that children, physically and mentally strong, cannot be raised in bad sanitary, physical and moral conditions. When society deals with its children as plastic material that will respond to the treatment they receive, and realizes that each normal child is capable of being trained in skill, intelligence and character so as to enable him to give good service to society, then will we discover our gain by the conservation of human brain and brawn.

Until such a system is instituted we must be content to pay the price of our neglect in saloons, busy courts of justice, expensive county jails, crowded penitentiaries, wasteful almshouses and an army of paupers. Dr. Charles Dudley Warner says that the criminal differs from the good citizen in the fact that the latter has contracted the habit of doing right and the former the habit of doing wrong. Z. R. Brockway says that prisoners need two things: "the habit of quick and accurate adjustment to good environment, and the habit of forethought." If these habits had been inculcated in early life by a system of education in the hands of masters, the presumption is that a large percentage of criminals would never have fallen into crime.

Intelligent legislative measures are needed on the question of alcoholism and criminality. John W. Griggs has said: "An extended experience of personal participation in legislation according to the American system (which we think is the best known) has led me to believe that there is no one thing in all the departments of government or business that is carried on with less scientific or orderly method than the making of laws." This is especially the case in our treatment of drunkards and criminals. All we do in both cases is partly to keep these anti-social elements within moderation and leave the sources, whence they are are fed, untouched. Our legislators have passed a law to buy blood-hounds * to aid our sheriffs in the discovery of criminals. If less crime were tolerated in high places and the laws were administered with Spartan impartiality and rigor against all classes of criminals in society, there would be no need of this resort to the savagery of slavery days. Professor Ferri says that 70 per cent. of discovered crimes goes unpunished, and of the murders committed only 2 per cent. of the murderers is executed. When legislation relative to criminals is made to harmonize with the conclusions of the science of penology, and when laws are framed by men who, by experience and scientific research, are thoroughly familiar with the subject, these anti-social elements, which prey upon the social body, will soon be eliminated.

The Legislature can also interfere to good purpose in the curse of alcoholism by limiting the number of saloons to the population, by punishing drunkenness, by holding the vendors of intoxicants responsible for the mischief they cause, and by placing the inveterate drunkard in safe-keeping where he will not afflict society with his nuisance and will not transmit his enervated will and depraved nature to innocent offspring.

Through Personal Effort.

Legislative measures at best are of uncertain value, and when once set in motion are, if their effect is mischievous, hard to correct. Individual efforts to elevate society are far more certain, and if they prove injurious to the common weal they can be more speedily adjusted to the conditions. If our people depended more on themselves and less upon the machine of government, we are sure that their social and industrial condition would be far better.

* This bill was vetoed by the Governor.

Sir Henry Maine said that "the unit of an ancient society was the family and of a modern society the individual." This is true but it is also equally true that man can only make progress by coöperation, and that society is the strongest which can bring its members to a perfect solidarity by the voluntary action of its units. Some one has lately said that the maximum of individualization will only be possible with a maximum of socialization. Much of the trouble of modern society arises because of the activity of the "devil of self-interest." If the units in society kept before them the collective welfare of the group and aimed so to regulate the activity of the social aggregate as to advance the interests of the whole, they would ultimately promote also their own interests. When individualization is synonymous with monopolization, then the sentiment of solidarity is scorched as trees are in a desert land. What peace can come when the working class and the capitalistic class arrange themselves in hostile camps and challenge each other to war? Our industrial leaders, with their clear vision and active brains, often smile at the mad surge of bitter polemic that engrosses the great mass of their employees, and regard the latter with contempt for wrestling with a truth that to the former has become trite, forgetting that the "common herd" lives by sentiment more than they do and that, under pseudo-ideas, they may be moved to action as strenuously as ever martyrs were who died for the truth. What we need is men of exalted ideas who believe in this "common herd," who realize that it is capable of a better fate than to be driven to the slaughter house. Danton once said: "If you suffer the poor to grow up as animals, they may chance to become beasts and rend you." We have seen a great mass of these workingmen raging and in its fury quenching the spark divine in man with a brutality that characterized savage beasts. Which is the better — to spread the light of truth among these men or do as a recent capitalist said: "Hands off, let's fight it out with guns."

Among the nations which came to the coronation of King Edward VII. of England, was a contingent of Fijian native constabulary that made a good appearance. A little over fifty years ago the last native king told Lord Rosmead who protested against his cannibalism: "It's all very well for you to talk in that way, you have plenty of beef in your country." The change has been effected by the touch of civilization. Man has always advanced by the masses following the lead of the upper classes. The Merchant Guilds of England in the thirteenth century rose under the leadership of strong men whose motto was, "Let all share the same lot." The daring and self-confidence of Queen Bess inspired her subjects, and the people did great things because they caught these qualities from their sovereign head. Cromwell said: "A few honest men are better than number. If you choose godly honest men to be captains of horse, honest men will follow them." We do not believe that the masses object to the concentration of riches in a few hands, if only the wealthy become leaders in the true sense of the word. What they object to is the foolish display of unbalanced plutocrats who for vain show spend more in one evening than is earned by the average workingman in a year. Millions spent on works of architecture and paintings may advance human culture and knowledge, but could not the culture and knowledge of men be better advanced by bringing the beautiful and the good within reach of the thousands of toilers into whose monotonous life very little that is beautiful and sublime ever enters? The fact that God "hath made of one blood all nations of men for to dwell on all the face of the earth" should never be lost sight of. Take the average among the poor and their possibility of advancement is equal to that of the average among the rich provided equal opportunities and advantages are given them. These people with tired hands and weary feet have modest longings in their hearts which are never satisfied, and are there more imperative demands upon the wealth of the sons of ease than to do something for these men whose lives are dead to the higher pleasures and refinements of civilization? Men who are willing to do this will not only benefit society but they will also justify the wealth which they enjoy and render it more secure. We are no pessimists; but suppose the party of discontent were to lose its faith in the righteousness of our present system of distribution of wealth and break loose, what would be the consequence? Would steel cars, buried vaults, marble walls, electric bells, etc., stay them from laying hands on private property? It is safer to appropriate a part of that wealth now, while it is day, to teach these "animals," lest the night come when the "beast" will break loose to rend the party of content.

The Le Play societies of France and the Positivists of England have raised the cry "Moralize the employer." They aim to teach him that he is a trustee of public as well as private interests. Our Commissioner of Labor, in speaking of the fathers of the factory system, says: "They were simple men of great intelligence, industry and enterprise. They have bequeathed the system to this age, with the imperfections incident to every human institution, and the task of harmonizing their innovation with existing institutions, and with the true spirit of righteousness, belongs really to the great employers of labor rather than to the professed teachers of morality." The same gentleman has also said of every captain of industry: "He is something more than a producer, he is an instrument of God for the uplifting of the race." The idea is too prevalent that the spiritual affairs of men are no concern of the employer; that he is only responsible for the wages they earn and, when these are paid, his responsibility ceases. Under this principle our moral decadence has come. The indifference of many of our leaders in the industrial world to the moral welfare of the men whose character they daily influence, has resulted in a lamentable indifference to morals in general. Man has never advanced save under social pressure, and when entrepreneurs lose sight of the sacred trust to mould the morals of those under them, the word of moralists and the precept of instructors have little effect. And not until the upper classes in our social hierarchy will again teach, and by direct example lead their employees to appreciate the Christian ideals, can we expect that regeneration of our society that will serve as a basis to raise a working population of sound mind and body.

The principle of vicarious sacrifice is in nature. Some of the cells of the brain are able to substitute themselves in the place of others and do their work. Sensitive nerves sometimes do the work of motor nerves. It is found also in society. The military power ever stands ready to sacrifice life and limb that the rest of the community may dwell in peace and safety. It has always been the principle which dominated the pioneers of science and art, religion and philosophy, who made possible the march of civilization. And is not this the very soul of the Nazerene's teaching in whose honor these churches which stud our valleys and mountains were built? And yet, with very rare exceptions, the leaders in our territory have not exercised it to the extent their business sense and general culture ought to suggest. Dr. Orello Cone has asked: "How much less would there be of unscrupulous gain, of the hot chase after riches, of the fever of demoralizing speculation, of the selfish greed that cares not whom it destroys, of the hard-hearted indifference to the poor, whose bodies and souls are converted into machines for producing wealth if men subordinated their activities to righteousness, honesty, kindness and fraternity?"

Some of the indicted may ask: "What can be done with such poor material?" We ask what have you tried to do? Some have built rooms and others have placed houses at the disposal of their employees, and, in some instances, these very appliances have fallen into the hands of the worst element in the towns and have become a curse and not a blessing. The experience of Octavia Hall, Edith Wright, Ellen Collins, Jane Addams, and a host of other consecrated men and women in our university and social settlements on both sides of the ocean, has established the fact that an old house, under the personal supervision of a competent person, gives better results than a model dwelling given in charge of the people themselves. "Arma virumque" made the epics of the old world and the same combination must be relied upon in these coal fields. Col. Boise pointed out one of the needs of our towns a decade ago, when he said: "Associations for the study and promotion of true Americanism ought to be organized in all the towns and cities where little attention is now paid to this or to the immigrant, save by the selfish politician with an eye for votes. Instruction in the use of the English language should have the first place among the purposes of these associations." None has been instituted.

Many operators have, long ago, observed the low social and moral conditions of employees living in company houses and yet the rentals have blinded them to the fact that one of the best agencies for the amelioration of these men would be the opportunity of buying that to which they have a just right, at a fair price and without being held under any obligation or asked for thanks. The saloon has devoured the substance of our people, reduced their industrial efficiency by debauching their bodies and debasing their souls, and yet in all our towns no substitutes have been planted and no efforts made, save in a few towns by a few clergymen, to reduce the number of these destroyers of youth and of chastity. Pauperism and crime have increased, the politicians have had a free course to run and become rich, and upon the thrifty and industrious the burdens of these anti-social elements have fallen, and not until lately have our public-spirited citizens of Schuylkill county moved to curb the extravagance of the one and the parasitism of the other. Will it not pay our capitalists, from considerations of economy only, to give attention to these questions in the anthracite coal regions? The strike of 1902 represented a total loss of over $100,000,000 because of a conflict which the intelligence of the twentieth century ought to have made impossible.

But all the wealth of these coal fields turned into philanthropic channels will not bring us relief unless the workers themselves resolve to raise their standard of social efficiency and industrial worth. The drinking done by mine employees * in these coal fields is appalling. It is the greatest enemy of organized labor, and if the Miners' Union will not try to

* In many of our towns for every cent spent on milk four are spent on intoxicants.

check it, the time will come when it will destroy the union. Labor leaders must feel and know that the greatest foe to the working classes is this accursed slavery to the "schooner." This is the reason that labor leaders in English trades unionism have become total abstainers and the temperance sentiment is working its way into the ranks of the Social Democrats of Germany. It should also be prominent in the Miners' Union if the organization is to remain permanent and its possibilities for good are to be fully realized.

Workingmen should also remember that the hierarchical principle is rooted in the very nature of things and that in no way possible can it be obliterated. Some men are made to govern, others to serve. The principle is due to differences of natural endowment which man is as powerless to obliterate as he is to explain. The inevitable consequence is that men are not equal and can never be. When William Morris read "Looking Backward" he felt a stifling sensation. No wonder, for by nature, his individuality wanted the arena of *laisser-faire* to play in, notwithstanding his advocacy of socialism. Who would want to live in a society whose historical type was violently disrupted by the intrusion of a scheme to level the inequalities established by the Maker of heaven and earth? It would be a hybrid composed equally of oriental despotism and a democracy void of all restraint. The Greek idea of life as a mountain guarded by lions is better. The brave and the wise tame the lions, the giants ascend the mountain and with strong arms tear the rocks and fill the gorges and, in spite of the thunderbolts, scale the heights and reach the summit amid the applause of the mass in the plain beneath. We need the brave, the wise and the giant to-day; their labors make ours lighter, their contributions lighten our burdens; and when wage earners will fully recognize the service of our leaders in the industrial army, they will cease to carp and bite and covet because they occupy a lower place in the social hierarchy.

The workers must also perceive the fallacy of the prevailing doctrine, that it is good for society when the wage earners spend every cent they earn. Superficial economists preach that the good of the community is served when men freely spend their money. No one will question the fact that thrift in the individual is virtuous, and can its practice by every unit in a group be mischievous to the community? All now-a-days are in a hurry to spend and make an ostentatious display of their wealth, so that an artificial air has permeated the working classes to the exclusion of the simplicity and the naturalness which once were their chief attraction.

> "Plain living and high thinking are no more.
> The homely beauty of the good old cause
> Is gone."

This is the reason of much of the social disquietude of the day. The wants of our working people increase much more rapidly than their wages, and the result is debt, anxiety and social impurity. It would be well for us to recall the words spoken of the greatest of American sages, Benjamin Franklin, about the middle of the last century: "The prevalence of habits of industry and economy, of foresight and thrift, of cautious calculation in the formation of plans, and energy and perseverance in the execution of them, and of the disposition to invest what is earned in substantial and enduring possessions, rather than expend it in brief pleasures or for purposes of idle show is due to an incalculable degree to the doings and sayings and history of this great examplar." If these virtues, which once belonged to the national character, were again incorporated in the mass of our working people, they would not turn their faces to the delusive dreams of communism and socialism.

Through the Church.

Masters of thought and expression tell us that our age is daily growing more materialistic. Dr. Cone says that the gigantic materialism of the times overtops and hides from the view of men the idealism of the New Testament. Others say that society has substituted political economy for religion, and has taken for its supreme aim to secure to the fullest possible degree the means of enjoyment. Dr. Brake has well said that this denotes "the organization of society from the point of view of the stomach-question. It has materialism as its presupposition and atheism as its necessary consequence." There is little wonder that men have come to this conclusion when an unbridled press sows broadcast the materialistic views of theorists who forgot all realities save what they can see or touch. When some of our teachers say that "supreme selfishness is presumed in all business transactions, and that to depend upon anything else is to build on sand," and "that the only means by which the condition of mankind ever has been or ever can be improved, is the utilization of the materials and the forces that exist in nature," there is little wonder that a thousand proud and insubordinate spirits echo the sentiments, knowing not whereof they speak. Society must ever live according to the laws of nature, but it is also true that the inexorable tribunal of history pronounces judgment on every nation that has tried to live contrary to the laws of morality. Let the sense of justice die out in the heart of the people and the foundations of society are shaken and all the relations of life are disjointed. Let false moral ideas and perversity possess the hearts of the people and the mischievous effect will be felt simultaneously in every sphere of human activity. We can never live and prosper unless the public preserves law and order, and there is no expedient yet found which can take the place of Christian ideals, upon which American society has in the past depended for its peace.

Carl S. Vrooman, writing last January in the *Outlook*, says of the popular university movement in France: "It is at bottom a religious movement. Scientific French agnosticism is becoming constructive, is trying to develop a religion. It has discovered that the cold truth of science is powerless except as warmed into life by the flame of a passionate altruism — that the scientific spirit and the Christ-like spirit are fundamentally and eternally necessary each to the other." The Father of our Country saw this when France discarded religion. He said in his farewell address: "Of all the dispositions and habits, which lead to political prosperity, religion and morality are indispensable supports. In vain would that man claim the tribute of patriotism who should labor to subvert these great pillars of human happiness, the firmest props of the duties of men and citizens. . . . Let us with caution indulge the supposition that morality can be maintained without religion. . . . Reason and experience both forbid us to expect that national morality can prevail in exclusion of religious principles." In the Pope's Cyclical letter of April, 1901, we find the same warning: "Society, in its foolhardy effort to escape from God, has rejected the divine order and revelation, and it is thus withdrawn from the salutary effects of Christianity, which is manifestly the most solid guarantee of order, the strongest bond of fraternity, and the inexhaustible source of public and private virtue." While the German Emperor in a striking speech on religion, in June, 1902, said: "It must not be forgotten that the empire was rooted in simplicity and the fear of God. I look to all priests and laymen to help me uphold religion among the people in order that the German name may preserve its health and strength." And Hon. Wayne MacVeagh, agreeing with these eminent personages, said that the Christian ideals should be constantly kept "before the minds of the plain people born in America, as well as before the minds of the hordes of untaught immigrants who are flocking to our shores from every quarter of the globe, (for) they have a tendency to soften their asperities, to lessen their animosities, and to encourage them to bear with greater patience the bitter and ever-growing contrast between the lives of idleness and luxury which we and those dear to us are privileged to lead, and the lives of labor and poverty which they and those equally dear to them are condemned to endure." These voices agree that religion is indispensable for the health and strength of the nation, and that the Christian type gives us the ethical ideals necessary to preserve these essential elements.

The earnest words of these leaders also suggest that the Christian ideals do not have the place they should have in the life of the people. Various reasons are assigned for this by apologists, but none denies the fact that the ethical teachings of

Jesus have not the power they once had over the hearts of men. It is not because there are not enough churches or a plentiful supply of well-trained clergymen — there is a superfluity of both. It is not because men are atheists and infidels, for the indictment is not true of our working classes. It is true, however, that the laity has lost faith in many doctrines which once were basic in the Christian faith and the rank and file of the clergy have failed thus far to adapt the ethical ideals of the Bible to the needs of the people of this generation. We believe also that the Protestant branch of the Christian church, in swinging so far away from the ritual and symbolism which formed so essential a part of Christian worship and in laying special emphasis on the sermon, appealed to the intellect at the expense of the heart and demanded of the masses greater intellectual power than they have time or inclination to give in order to make divine service a success. The *οἱ πολοί* do not live so much in the head as in the heart, and in the barren walls and plain pulpits of most Protestant churches there is little to inspire the heart to devotion. If any powers of the soul are more exercised than others in divine worship, they are the sentiments, and in Protestant churches, especially under the ministration of clergymen of mediocre ability, the feelings are very slightly touched. The churches where ritual and symbolism are used are the most flourishing and have in them the promise of long life, and if by the restoration of these aids greater interest could be aroused in the hearts of the worshipers in Protestant churches it would seem advisable to restore them. Surely this course would be far more commendable than the sensationalism resorted to by many preachers to "draw the crowd." The church will be in the world as long as men feel the badness of sin and the goodness of virtue ; it will be maintained as long as love is more attractive than hate, but the danger lies in the fact that it will not be to the people of the twentieth century what it has been in the history of every nation in Christendom — a restraining and purifying power, a regenerating and inspiring force, a unifying and uplifting factor in society. And when the church shall again mean all this to society then will its social service increase.

The great need here again is men. There was a type of men in days of old who wielded mighty power with the people. Cuthbert, the missionary, who said: "Never did man die of hunger who served God faithfully," was one of them. Our age, given to stomach worship, says that fasting and asceticism are not needed; half the Christian church has thrown them away and the other half retains the semblance. But are there not spirits in the twentieth century which cannot be cast out save by "prayer and fasting"? Ruskin has told us that no good will ever be done in teaching the eternal truth at the rate of a penny a line. The old teacher, Edmund Rich, wielded a tremendous influence, although he used to say, when money was offered him for teaching, "ashes to ashes and dust to dust," and threw the money away. But we live in a commercial age when the dignity of the profession will not allow talent to be wasted in gratuitous service, and yet the sheep wander without shepherds whose voice they will hear. There once was a man who said: "The zeal of thine house hath eaten me up." Never was there any great cause in the history of the world which succeeded without zeal, and the man who is willing to give himself a living sacrifice on the altar of faithful service for the people will not lose his reward.

We also want men who will, by careful study of the Bible, make it a living book to the men and women of the present day. There is no other book that can take its place as an exhibit of the operations of the divine spirit upon the minds of men, and if by scholarly and scientific research, the men of to-day can understand the conditions under which these manifestations were possible in the development of the human mind, the social service they will render to society will be inestimable.

In all human affairs economic considerations have had the first place in the minds of men, and the periods of disaster and ruin which have fallen upon nations have been those when the soul sang its sweetest, thought its deepest and prayed its most fervent prayers. We live under very different economic conditions from those under which Israel lived, but if our intellectual life kept pace with our industrial, we would provide a place in our system for the moral and the religious factor upon which, as far as the inexorable judgment of history shows, every nation, regardless of its stage in civilization, has depended. The Lochrians always left a vacant place in their charging ranks for the spirit of Ajax Oileus. If modern civilization, in its industrial advancement, had been careful to provide a place for the divine Spirit in its ranks, the ominous developments now seen on all sides would not threaten us. Are we to be taught the great lesson of history again by deterioration and retrogression? The only way to avoid it is by the reviving of the prophetic spirit which will speak with authority to the men of to-day as it did to Israel of old.

For this purpose it is time that the best intellects of the Church should come together and devise means to this end. It is true that the Spirit goes where it listeth, but man has much to do in preparing the way for His coming. Professor Paul Haupt asked the other day "if there is any difference between Catholic and Protestant mathematics, or between Christian and Jewish physics, or between Episcopalian and Presbyterian chemistry," and concludes that when the Bible is scientifically studied "only one interpretation can be correct." On the questions pretaining to Biblical exegeses there will be differences such as have divided scientists, but it would seem reasonable to expect that intelligent and patriotic men, in view of the exigencies of the times, should find a common standing ground as to the best method of teaching the principles of morality and religion to the rising generation. This is not satisfactorily done by the Sunday-school and the home. Anyone familiar with the work done in the Sunday-schools and homes of our communities knows whereof we speak. We are fast coming to the condition that even among those who still profess to adhere to the Church, knowledge of the Bible is nil, familiarity with the fundamental principles of Christianity is most meager, and reverence for sacred things and authority is fast dying out. Under such conditions is it strange that religious hobbies prevail and gross superstitions flourish? There is nothing in the hearts of these youths to resist the encroachments of error and folly, and the religious sentiments, not having substantial food to feed on, feed on the vagaries of cunning men. Against this tendency, so apparent that he who runs may read, ought not the leaders of religious thought to come together and devise an effective scheme for the religious instruction of the youth which will strengthen them in the knowledge of the Scripture and ground them in the faith of their fathers? There may come a day when the souls of men will reach a stage of development that will enable them to walk in the light of divine truth and be wholly independent of both Bible and Church. We are far from that stage to-day. And those men who know how man has advanced in the quest after God must feel that the means in common use among Christians to promote the religious life cannot be dispensed with. It is difficult to conceive of a greater disaster to society than violently to wrench the traditional system of religious instruction from the minds and hearts of the people. In this, as in all else, the need is to clarify, to strengthen, and to progress, and if done by the coöperative action of the leaders in the religious world, we may confidently expect a healing influence to come from the Church of God that will remedy many of the diseases from which we now suffer.

INDEX.